Aristotle on the Matter of Form

Cycles

Series Editors
Andrew LaZella, The University of Scranton
Richard A. Lee Jr., DePaul University

Critical perspectives on the history of philosophy, from Ancient Greece to the nineteenth century, expanding traditional approaches to Western Philosophy

Cycles shows how the history of philosophy is a living tradition, influencing and determining our thinking in the present while maintaining its deep and rich roots in the past.

Editorial Advisory Board
Jason Aleksander, Robert Bernasconi, Emanuela Bianchi, Sara Brill, Peter Casarella, Idit Dobbs-Weinstein, Dan Selzer, Kristi Sweet

Books available
Aristotle on the Matter of Form: A Feminist Metaphysics of Generation
Adriel M. Trott

Aristotle on the Matter of Form

A Feminist Metaphysics of Generation

Adriel M. Trott

EDINBURGH
University Press

Edinburgh University Press is one of the leading university presses in the UK. We publish academic books and journals in our selected subject areas across the humanities and social sciences, combining cutting-edge scholarship with high editorial and production values to produce academic works of lasting importance. For more information visit our website: edinburghuniversitypress.com

© Adriel M. Trott, 2019

Edinburgh University Press Ltd
The Tun – Holyrood Road, 12(2f) Jackson's Entry, Edinburgh EH8 8PJ

Typeset in 11/13 Foundry Sans and Foundry Old Style
IDSUK (DataConnection) Ltd

A CIP record for this book is available from the British Library

ISBN 978 1 4744 5522 0 (hardback)
ISBN 978 1 4744 5525 1 (webready PDF)
ISBN 978 1 4744 5524 4 (epub)

The right of Adriel M. Trott to be identified as the author of this work has been asserted in accordance with the Copyright, Designs and Patents Act 1988, and the Copyright and Related Rights Regulations 2003 (SI No. 2498).

Every realm of nature is marvelous: and as Heraclitus, when the strangers who came to visit him found him warming himself at the furnace in the kitchen and hesitated to go in, is reported to have bidden them not to be afraid to enter, as even in that kitchen divinities were present, so we should venture on the stuff of every kind of animal without distaste; for each and all will reveal to us something natural and something beautiful.
– Aristotle, *Parts of Animals* 645a17–23

οὐ γάρ πώ τις ἑὸν γόνον αὐτὸς ἀνέγνω.
No one truly knows his own begetting.
– Homer, *Odyssey* I.215

Contents

Acknowledgements	ix
Introduction: Aristotle and the History of Sex	1
Method	8
Historical Approaches to Sexual Difference	13
One-Sex and Two-Sex Models	16
Aristotle and the One-Sex and Two-Sex Models	20
Material Evidence	23
1. Feminist Critics of Aristotle's Biology and Metaphysics	27
The Feminist Disputes	27
Feminist Critics Who Say Aristotle's Biology is Sexist	28
Critics Who Say Aristotle's Biology is Not Sexist	30
Feminist Critics and Defenders of Aristotle's Metaphysics	37
Conclusion	48
2. Disputes over the Material Contribution of Semen	50
Robust Material View versus Restricted Material View	52
Material Necessity and Teleology	62
Aristotle and Plato on Form	64
Conclusion	77
3. Aristotle on Material	79
Beyond Prime Matter	79
Physics I.7: Change and the Substratum	82
Metaphysics VII.3: Substance as Substratum	90
De Caelo IV.5: The Elements in Common	92
How Matter is Many	97
On Generation and Corruption	102
Matter in Natural Substances: Generation as a Problem	109

4. The Feminine and the Elemental in Greek Myth, Medicine and
   Early Philosophy   120
      Gender, Generation and the Gods   121
      The Elemental among the Pre-Socratics   131
      Hippocratics, Elemental Powers and Gender   137
      Conclusion   140

5. Semen, Menses, Blood: Material in Generation   143
      Concoction as Mastery of the Moisture   146
      Internalised Heat: From Blood to Semen   154
      Material and Teleology in Aristotle's *Meteorology* 159
      How Semen is a Causal Part   165
      Material Composition of Semen   176
      Sources of Vital Heat: Stomach, Sun and Earth   181

6. Sex Differentiation, Inheritance and the Meaning of Form
   in Generation   187
      'The female is, as it were, deformed'   187
      How Sex Differentiation Explains the Work of Form and Matter 191
      How Inherited Traits Explain the Role of Form   197
      Contradiction and Contrariety   204

7. Craft and Other Metaphors   212
      Craft Analogy Explains Why Generation Needs Males and
         Why It Happens in Females   212
      Craft Analogy Explains How Semen Actualises Form 219
      Other Images and Metaphors   227

Conclusion: On Material in Aristotle's Biology   236
   Revisiting the One-Sex and Two-Sex Models   238
   The Möbius Strip   240

Bibliography   242
Index Locorum   255
General Index   260

Acknowledgements

The moving cause of this book's generation was two encounters with feminist scholars engaging conceptions of gender in ancient Greek philosophy. The first was with Emanuela Bianchi, who challenged me to consider whether my account of nature in Aristotle's *Politics* was supported by his account of nature in the biological works. No disagreement in my scholarly life has been as productive to me as ours. The second was with Brooke Holmes's exemplary study, *Gender: Antiquity and Its Legacy*, in which Holmes joins many other feminists when she charges Plato and Aristotle with grounding philosophy 'on the denigration of a feminine principle identified with materiality, bodies and emotions', in a way that should lead us to 'care about the ancients not because they're alien but because we're still dealing with the "seamier legacies" of antiquity, among which patriarchal bias and systemic misogyny rank high'.[1] Holmes's careful multivalent analysis of portrayals of sexual difference in ancient Greek literature, medicine and philosophy led me to believe that Aristotle's texts were equally fertile ground for a reading that draws attention to the tensions in the text to show the interrelated Möbius strip relation of apparently opposed positions.

I am grateful to Walter Brogan and Sean D. Kirkland, who both read and gave generous and thoughtful comments on the manuscript. L. Aryeh Kosman and Rebecca Goldner offered supportive feedback on the project early in its development. During the course of this project, I have benefited from the intellectual friendship and conversation occasioned by the Great Lakes Colleges Association (GLCA) Ancient Philosophy Teaching and Research Collaborative Initiative (APTRCI) with my GLCA colleagues Lewis Trelawny-Cassity and Kevin Miles. Our conversations and friendship have sharpened my thinking.

---

[1] Holmes, *Gender: Antiquity and Its Legacy*, p. 7.

The Pennsylvania Circle of Ancient Philosophy's invitation to keynote in 2014 was the occasion of my first presentation of the ideas at the core of this book. The invitation from the organisers, Laura McMahon and Hande Kesgin, motivated me to begin this work in earnest. My colleagues at Wabash College including Cheryl Hughes, Glen Helman (who should be credited with the idea for the cover) and Joseph W. and Leslie P. Day offered critical feedback on this work when it was still in its early stages. The students and faculty of the philosophy department at the University of South Florida were a constructive audience. Danielle Layne and the rest of the faculty at the Philosophy Department at Gonzaga University graciously hosted me for the Annual Rukavina Lecture in Philosophy, where they helped me see some of the obstacles that I would need to overcome to make my case. Elly Pirocacos and the philosophy faculty and students at the American College of Greece and the GLCA APTRCI have my appreciation for the occasion and the means to share this work in Greece. I had a productive discussion of some of the central ideas in this work at a meeting of the Society for Ancient Greek Philosophy in New York in October 2016. Lee McBride and the Philosophy Department at the College of Wooster and Barbara Fultner and the Philosophy Department and Women's and Gender Studies Program at Denison University and their students offered the happy occasion to present the most final version of this project, and I am grateful for their hospitality and their enthusiastic reception.

The Ancient Philosophy Society (APS) has been a haven for conversation about this work and related work that has sparked my thinking. In particular, Eric Sanday at the University of Kentucky and Jeremy Bell and Marta Jimenez at Emory University offered the occasions at the two meetings they organised for me to share work from this project, and Holly Moore offered commentary that contributed to a sharper structuring of this project. Sometimes you need other scholars who might not work on the same material but share similar impulses in reading ancient Greek philosophy to remind you of the principal investments of your own work. In this respect, I am grateful for conversations I had with Mitch Miller and Jill Frank at these respective meetings, who in different ways reminded me of my impulse behind this project to show the unity of Aristotle's causes and thus of Aristotle's sense of nature.

Abraham Jacob Greenstine and Ryan J. Johnson supported the development of this project by inviting me to contribute to their anthology, *Contemporary Encounters with Ancient Metaphysics* (Edinburgh University Press, 2017). My contribution to that volume contains embryonic versions of arguments found in the Introduction, Chapter Five and Chapter Six of this monograph.

Carol Macdonald, my editor, offered wisdom and direction in bringing this project to publication. A sabbatical leave from Wabash College granted me the time to write the bulk of the manuscript for this book.

Finally, for his unwavering support of me, my work, and this project, I am overwhelmed with gratitude to Jeff Gower, with whom I am irreducibly intertwined.

Introduction: Aristotle and the History of Sex

I argue in *Aristotle on the Nature of Community* that Aristotle's account of nature as an internal source of movement offers possibilities for thinking political life that are neither restrictive nor exclusive. In this emergent sense, nature does not impose form on material but form arises from within a natural thing.[1] I make the case that an internal source of movement – ἀρχὴ κινήσεως – is what Aristotle means when he describes political life as natural – it emerges from within our being together concerned with what is good and just. In my efforts to consider how a conception of the political can recognise the nonideal tendencies toward exclusion that constitute actually existing communities, I argue that Aristotle offers a conception of community that institutes the concern with how it falls short. The emergent conception of nature, based on the definition from *Physics* II.1 as ἀρχὴ κινήσεως, makes the end and achieving the end a constant concern for the community because the form and the final cause coincide.[2] That is to say, in this emergent sense of nature, natural things work from the form internal to them to actualise their being, which is their final cause and what it is for them to be what they are. In political life, this concern for fulfilling our end from within ourselves requires vigilance and can be prompted by the continued question of the relation of the community to itself and to what appears to be outside of itself. The emergent conception of nature situates that effort within the being that is natural, always working on becoming that which it is or, as in political life, claims to be.

Accounts of nature that oppose nature to reason and craft, to external intervention, make of nature that which is the given orderable stuff of the world that

---

[1] Adriel M. Trott, *Aristotle on the Nature of Community*. For this account of nature as principle of movement by contrast to the views of nature as mere givenness or the hand of an artisan, see pp. 19–41.

[2] *Phys.* 192b14–22.

needs to be structured and by so doing divide the world between what orders and arranges and what needs to be ordered and arranged. This view of the world construes what is as divided between the natural and the reasoned, determined, and structured. In projects of making, the form in the mind of the artist is imposed on material designated as suitable for the making by its more natural processes that produce a stuff for the higher end of τέχνη, craft. In τέχνη what is most complete is finished and determined, and when finished is separated from the governing principle that brings it into being. The form in the mind of the artisan ceases to be at work. It has a form insofar as it stands as a shape, but not as some internal principle working on making it be what it is.

Dividing the world between φύσις, nature, and τέχνη, craft, leads to views of nature that internalise this opposition and as a result cast nature in a τέχνη framework.[3] Nature itself is then divided between what is orderable and what orders, between what is unstructured and what structures, between what is arrangeable and what arranges. Aristotle takes aim against Antiphon in *Physics* II for supposing that the wood is what most is in the bed analogy Antiphon employs to claim that nature is most essentially material. Antiphon reasons that if you planted a bed, a tree, the source of the wood out of which the bed is made, would grow. Antiphon concludes that nature is more that which is arranged, because he thinks that an object of craft is an interchangeable analogy to natural things. While Antiphon is right to look at the naturalness of the bed in what would grow up from it if planted, he fails to see that what would grow is different from the artificial bed because the tree has its own principle of ordering itself to bring it into the complete work of being a tree. This view of nature makes the natural being only incidentally related to the ἀρχὴ κινήσεως from two perspectives. Nature is only incidentally related to its governing principle when it is viewed as the stuff that is just there, as it seems to be for Antiphon, where material is viewed as indeterminate and waiting to be arranged. It remains only incidentally related to its governing principle when the ἀρχὴ κινήσεως is viewed as an internal craftsperson who orders the material stuff and aims to fix and determine the indeterminate material that exists for form to do its work. Nature is seen as a bare stuff available for production when nature

---

[3] Martin Heidegger develops this distinction between Aristotle's account of φύσις and a τέχνη model in 'On the Essence and Concept of Φύσις in Aristotle's *Physics* B, 1'. For more on Aristotle's account of τέχνη as a production whose work and end is beyond or outside it in such a way that shows how τέχνη is distinct and separable from which it produces by contrast to φύσις, see Heidegger's lectures on *Nicomachean Ethics* VI in his lectures on Plato's *Sophist*, pp. 27-40, 62-4, 71, 98. For more on φύσις as contrasted to τέχνη, see Claudia Baracchi, *Aristotle's Ethics as First Philosophy*, pp.183-6, 186n39.

occupies the side of nature against τέχνη or it is viewed as the craftsperson when nature occupies the position of τέχνη in the shape of the artisan or form. Viewed in this way, nature is never unified, but the source of a further division in the world. It has no impulse to change in itself, but must always be moved from something other than itself. In this sense part of it is always imposing and part of it is always being imposed upon. This condition would make nature, as in τέχνη, never wholly for itself. An art is never for the sake of itself – to increase itself – but for something outside of it, as doctoring is for health.

While Aristotle has been saddled with readings of nature that seem to follow the technical model, his definition of φύσις as ἀρχὴ κινήσεως καὶ μεταβολῆς, the principle or source of movement and transformation, would resist the making view of nature and the oppositional contraries that constitute it.[4] On Aristotle's account, nature comes to be from within itself. Nature holds within itself its own principles of movement and actualisation. The activity of nature is for more nature, not for something otherwise that does not organise its own activitiy. The self-governing of nature is both because it aims toward more nature and because it works from within itself to fulfil its end. To do this work of moving from within itself, nature has to involve a co-dependence of ὕλη or matter and εἶδος or form.

While Heidegger argues that γενέσις is the movement in which nature most shows its character,[5] Aristotle's account of generation in the biological works would seem to challenge my case that Aristotle defines nature in this way because at the micro-level of generation it looks as though form as semen is imposed on material menses. In one characteristic passage, Aristotle writes:

> [T]he female always provides the material, the male that which fashions it, for this is the power that we say they each possess, and this is what it is for them to be male and female. Thus while it is necessary for the female to provide a body and a material mass, it is not necessary for the male, because it is not within what is produced that the tools or the maker exist. While the body is from the female, it is the soul that is from the male, for the soul is the substance of a particular body.[6]

In this passage, Aristotle sets up a metaphysical binary between form and matter, soul and body, male and female. The binary hierarchises the male

---

[4] *Phys.* 200b12.
[5] Heidegger, 'On the Essence and Concept of Φύσις in Aristotle's *Physics* B, 1', p. 220 and *Basic Concepts of Aristotelian Philosophy*, p. 153.
[6] *GA* 738b20-6.

contribution of form and soul, which is described as the substance of the body. Toward the end of this same chapter, Aristotle offers more specifics:

> The female (τὸ θῆλυ), then, provides matter (ὕλην), the male (τὸ ἄρρεν) the principle of motion (τὴν δ'ἀρχὴν τῆς κινήσεως). And as the products of art (τὰ ὑπὸ τῆς τέχνης) are made by means of the tools (διὰ ὀργάνων) [of the artist], or to put it more truly by means of their movement (διὰ τῆς κινήσεως), and this is the activity (ἐνέργεια) of the art (τέχνης), and the art (τέχνη) is the form (μορφὴ) of what is made in something else, so is it with the power of the nutritive soul (θρεπτικῆς ψυχῆς δύναμις). As later on in the case of mature animals and plants this soul causes growth from the nutriment, using heat and cold as its tools (ὀργάνοις) (for in these is the movement of the soul . . .).[7]

This passage would seem to have all of the elements of a τέχνη model of nature: the form as an artist who works on the inert matter to bring it into being, externally imposing itself. In a later passage, Aristotle again seems to describe the τέχνη model of nature, this time in such a way that defines the female and matter by contrast to the male and form in terms of a lack:

> But the male and female are distinguished by a certain capacity (δυνάμει) and incapacity (ἀδυναμίᾳ). (For the male is that which can concoct (δυνάμενον πέττειν) and form and discharge a semen (σπέρμα) carrying with it the principle (τὴν ἀρχὴν) of form (τοῦ εἴδους) – by 'principle'(ἀρχὴν) I do not mean a material ~~principle~~ (ὕλης) out of which comes into being an offspring resembling the parent, but I mean the first moving cause (κινοῦσαν πρώτην), whether it have power to act as such in the thing itself or in something else – but the female (θῆλυ) is that which receives (δεχόμενον) ~~semen~~, but cannot form (συνιστάναι) it or discharge (ἐκκρίνειν) it).[8]

In this passage, the formal contribution of the male is conceived chiefly as a moving cause that shapes what the female offers because she lacks the capacity that the male has. Aristotle emphatically insists that the source or principle is form which works by moving and that the female receives and is moved by

---

[7] GA 740b24-31.
[8] GA 765b8-15. I present the translation here from A. Platt (Barnes) with strikethroughs where words that are not found in the Greek are interpolated in the English. I continue this practice especially in the second half of the book, where careful attention is paid to these passages.

this principle. In discussion of how resemblance and sex differentiation occur, Aristotle again seems to sketch out this τέχνη model:

> The reason why the movements relapse (λύεσθαι) is this. The agent (τὸ ποιοῦν) is itself (πάσχειν) acted upon by that on which it acts (ὑπὸ τοῦ πάσχοντος); thus that which cuts is blunted by that which is cut by it, that which heats is cooled by that which is heated by it, and in general the moving cause (τὸ κινοῦν) (except in the case of the first cause of all) does itself receive (ἀντικινεῖταί) some motion (τινα κίνεσιν) in return; e.g. what pushes is itself in a way pushed again and what crushes is itself crushed again.[9]

This passage describes an active agent who is the moving cause and a passive patient. While it seems to allow the passive patient some resistance, it describes it as that to which things are done by contrast to that which does things. On their face, these passages seem to suggest that nature as an ἀρχή κινήσεως would not in every case move from within, but be imposed on matter from without in a τέχνη model of nature. Such a view divides form from matter and privileges form, which carries political consequences of exclusion (matter left unformed – or those more associated with matter – would not have any claim of their own) as well as gender consequences of a hierarchical binary that privileges the forming pole of the binary.

Moreover, these passages point to the way that Aristotle's biology makes explicit the gendered nature of the distinction between form and matter. Form is associated with semen, matter with menses. On a τέχνη model of generation, the male principle of form brings life and meaningfulness; the female principle of matter has no meaning of its own but needs form to appear in the world. Aristotle appears responsible for the ways we continue to think gender difference today, where a binary divides between men who give shape and meaning to the world and women who offer the material underpinnings of it, and the distinction between them is strongly policed. As a principal source of this distinction, Aristotle is a fruitful source for thinking more carefully how to challenge this way of thinking of sex and gender difference, and relatedly, of thinking of the relation of form and material in hylomorphism.

In the twentieth century, feminists unsuture gender from sex, arguing that it was not the body that dictated separate roles to men and women, but rather cultural demands and expectations. As Beauvoir writes, 'One is not

---

[9] *GA* 768b15–20.

born, but rather becomes a woman.' This strategy liberated women from culturally designated roles disguised as natural. The unfortunate consequence of this strategy was to re-inscribe the natural at the site of biological sex, making sex what is given and material, the element that cannot be altered. By shifting the discussion to gender, the notion that sexual difference, the difference at the level of the body, is 'natural' is left unchallenged. Instead of challenging the difference altogether, articulating it at the level of culture places sexual difference at the level of what is unalterable and given, the biological body. As Elizabeth Grosz points out, this strategy contributes to a 'presumption of a base/superstructure model in which biology provides a self-contained "natural" base and ideology provides a dependent parasitic "second story" which can be added – or not – leaving the base more or less as is'.[10]

This model appears to follow from an Aristotelian view of form as what changes the world and gives it meaning, and of matter as what remains the same. The natural and given is opposed to what is capable of change. Form is associated with the cultural, historical and rational soul, and matter with the natural perduring body. Sexual difference is what is natural and material. As such, it limits and prescribes what is possible. Further, when the material body becomes the truth of difference, those persons more associated with materiality become more destined by their bodies than those who are perceived to be capable of transcending the limitations of their bodies and of actualising their rational souls.

Judith Butler argues that Luce Irigaray falls prey to this problem of inscribing sexual difference at some more fundamental level of givenness in her reading of the history of philosophy. Irigaray accuses philosophers from Plato to Heidegger of basing their philosophical systems on a forgotten outside. In her reading of Aristotle, this forgotten outside is matter or nature as givenness. Irigaray argues that Aristotle bases his conception of the body, the sexed body, on a more primordial forgotten material associated with the mother.[11] As Butler reads her, Irigaray maintains that 'the feminine is cast outside the form/matter and universal/particular binarisms' as the 'permanent and unchangeable condition of both'.[12] Irigaray's strategy points to the power of mother-matter by showing how the whole project is grounded in some more primordial matter that allows form to appear. Following Irigaray, Emanuela Bianchi argues that Aristotle treats material both as pure δύναμις, potentiality, for form's work and as ἀδυνάμια, impotentiality, having no capacity of its own. These dual roles of material show it to be what Bianchi calls a repressed and aleatory

---

[10] Elizabeth Grosz, *Volatile Bodies: Toward a Corporeal Feminism*, p. 21.
[11] Luce Irigaray, 'How to Conceive (of) a Girl', in *Speculum of the Other Woman*, p. 164.
[12] Judith Butler, *Bodies That Matter*, p. 16.

principle that, as repressed and other, returns to trouble the work of form.[13] This double role of material is what makes sex both the unchanging given ground and that which needs to be enforced.

Butler worries that Irigaray might be situating material and the maternal in this exterior position that makes it the original given in the same way the gender/sex binary divides culture from nature to challenge what has been construed as natural, but ends up more deeply entrenching nature on the other side of the binary. Culture and the capacity to change seems itself to be the formal principle against the biological material given. To Butler, Irigaray seems to establish material as the 'sign of irreducibility' that bears and supports not only form but the construction of culture in her efforts to give the feminine its own power and definition. Such a view seems to further isolate material and produce a division between what is changeable, cultural and what is natural – a true original ground.[14]

Arguing against the view that material itself exists as a pre-constructed basis, Butler observes that Irigaray's strategy attempts to establish material as a critical ground from which 'to verify a set of injuries or violations only to find that *matter itself is founded through a set of violations*'.[15] For Butler, the very natural givenness of material is constructed. Irigaray's reading of Aristotle's natural science poses a similar problem that arises when political community forms from opposing reason to nature or hierarchising reason over nature: a fundamental exclusion results from the opposition. For Irigaray, the construction of the feminine outside is due to an opposition and subordination of matter in relation to form. Any attempt to recover a robust sense of material in Aristotle thus seems subjected to the problems that Butler finds in Irigaray's reading – either we posit material as this constitutive outside that is other than and distinct from form, the 'true ground' that is most genuinely given, the real nature, the ur-maternal, or we describe it as always already subjected to form, or both.

Butler maintains, 'If matter never appears without its *schema*' in Aristotle, 'that means that it only appears under a certain grammatical form and that the principle of its recognisability, its characteristic gesture or usual dress, is indissoluble from what constitutes matter'.[16] The danger of positing separate principles of form and matter is that material becomes legible solely under

---

[13] Emanuela Bianchi, *The Feminine Symptom: Aleatory Matter in the Aristotelian Cosmos*, esp. Chapter Six, 'Sexual Difference in Potentiality and Actuality', pp. 183–222. See Gayle Salamon on Irigaray's reading of material in *Physics* IV, 'Sameness, Alterity, Flesh: Luce Irigaray and the Place of Sexual Undecideability'.
[14] Butler, *Bodies That Matter*, p. 4.
[15] Ibid. p. 5.
[16] Ibid. p. 8.

the guise of form, which then stacks the deck in favour of form, making it impossible to argue against the conception of material as stuff that needs form. This account of material's need for form relies on an account of the material principle as already separable and distinct, *other*, and needing form. But if, as I shall argue in this book, form in Aristotle's biology comes to be in an intensification of heat in material, an intensification that also occurs in material but to a lesser degree, then there is both a necessary contribution of matter in constituting form and a distinct character to material that is irreducible to form. Both semen and menses depend on material processes to become what they are. As Brooke Holmes writes, 'The idea that matter has a (feminine) gender becomes, in Butler's hands, a myth to be exploded.'[17] To explode the myth of the femininity of matter without denigrating either femininity or matter is to reconsider anew what material as such is.

Method

This reading of Aristotle grants the position of those feminist critics who argue that a strong distinction between form and material establishes a normative hierarchical relationship between the two principles. A sympathetic reading of Aristotle, therefore, must show not only that he gives more of a role to the feminine contribution, but that the distinction between form and material is not as severely drawn. Elizabeth Grosz provides a useful image of a Möbius strip to explain the relationship between form and matter that involves both difference and interdependence, not only in forming substance, but in being form and material. Clarifying what is most devastating about the feminist critiques of Aristotle's metaphysics sets the stage to explain how the biology resists the received reading of Aristotle's physics and metaphysics. Against those like Jonathan Barnes who distinguish between more and less philosophical treatises in Aristotle's corpus and argue that Aristotle's biological treatises are of merely historical and not philosophical interest,[18] this reading shows how readings of Aristotle's distinction between form and matter continue to affect the way we understand gender difference. New strategies for thinking that distinction offer new conceptual apparatus for thinking gender.

I do not think that Aristotle, according to Robert Mayhew's definition of ideology, is explicitly trying to rationalise and thereby dishonestly enforce unfair power relations through philosophical and scientific arguments.[19] Daryl

---

[17] Brooke Holmes, *Gender: Antiquity and Its Legacy*, p. 67.
[18] Jonathan Barnes, *Aristotle: A Very Short Introduction*, p. 139.
[19] Robert Mayhew, *The Female in Aristotle's Biology*, pp. 2–6.

McGowan Tress argues that Aristotle cannot be judged on the basis of equal value of male and female because Aristotle does not develop this imbalance out of misogyny or sexism but because of the particular problems that he is addressing.[20] My concern is not with Aristotle's intentions, because I do not consider sexism to be a matter of personal bias but of structures that reproduce and enforce inequality between the sexes. I do think, following G. E. R. Lloyd's definition of ideology, which as Sophia M. Connell notes would seem to make every thinker almost inevitably ideological, that Aristotle accepts and incorporates cultural and social norms into his theory.[21] The problem is not ultimately those cultural and social norms, since we could argue that they have changed, but the way that these norms became attached to a metaphysics that influences a tradition, which even when it aims to strip inequality from its metaphysics still includes the remnants of the dichotomous association of form with male and matter with female.

Malcolm Schofield's definition of ideology as arguments or doctrines that form a system of bias against specific groups would suggest that Aristotle is culpable.[22] The point is not so much Aristotle's culpability, but that his account operates intentionally or not to form a system of hierarchy that privileges male form over female material. My approach follows the efforts of those like Tress, Connell and Mayhew who situate and contextualise Aristotle's arguments in order to understand their significance in terms of the arguments and thinkers to which he is responding. These thinkers show various ways that Aristotle in his context gives more of a role to the female in reproduction than the Hippocratics, and that Aristotle does not invent the association of the feminine contribution with moisture, but rather complicates and develops a sense of moisture that makes it more than mere stuff. While this approach allows for an analysis of Aristotle in the terms of the cultural milieu in which he was thinking, it also recognises that the association of male with form and female with material has significant normative consequences for conceptualising gender and metaphysics. Making those consequences explicit lays the groundwork for arguing that Aristotle's account makes semen and menses and form and matter more interdependent and less strictly divided than the received reading of the difference between form and matter indicates.

---

[20] Daryl McGowan Tress, 'The Metaphysical Science of Aristotle's *Generation of Animals* and Its Feminist Critics', p. 310.

[21] G. E. R. Lloyd, *Science, Folklore, Ideology: Studies in the Life Sciences in Ancient Greece*, p. 20. Sophia M. Connell, *Aristotle on Female Animals*, p. 44.

[22] Malcolm Schofield, 'Ideology and Philosophy in Aristotle's Theory of Slavery'. See Connell, *Aristotle on Female Animals*, pp. 48–9 and 'Conclusion', and Sarah Borden Sharkey, *An Aristotelian Feminism*, p. 82.

One point of departure in this monograph is Luce Irigaray's critique of Aristotle. Butler's critique of Irigaray is warranted but it does not amount to a dismissal of the most damning of Irigaray's criticisms, which is that the female material is considered merely for the sake of the male form, which inaugurates the tradition in the history of philosophy of making the female for another and the male for itself. While the female animal is clearly an actualised substance, the association of her role in generation with a kind of passivity that characterises material subordinates her role to the forming and actualising role of the male. Whether the material contribution is passive or a lack is a matter of dispute that this book will consider at length. That it is a matter of dispute does not mean that Irigaray is offering a nonsensical reading of Aristotle, who analogises between the female and matter at the end of *Physics* I and who describes the female in language similar to how he describes matter – as passive, as potential, as receptive, and so forth. Some commentators distinguish between the female animal and the female principle to respond to this critique, but it is not clear that such a distinction undoes the association of the female animal with this privative position.

Irigaray argues that the history of philosophy functions much the same way that form and matter function in Aristotle: the form of thought achieves its fulfilment in material that is suppressed in its own identity for the sake of the form's actuality. She makes this argument through a project of reading she calls 'mimicry' in which she points to places in the text where something must be overcome in order for true thinking or true being to become possible. That something must be overcome makes of the thing that must be overcome the foundational possibility of the so-called true thinking or true being.[23] For example, in 'How to Conceive (of) a Girl', Irigaray draws attention to the moves that Aristotle makes, retelling them from the point of view of material outside Aristotle's story of progress and of becoming rather than form and being. This perspective teases out the way that Aristotle, or at least, the history of the reception of Aristotle, decided for a certain account – both of a biology and a metaphysics – that was not inevitable. On this account, Aristotle needs to put the excessive power of the elements in their proper, and thus controllable, place in order to make the generative work of the mother manageable and knowable. Irigaray's analysis of generation weaves through the claims that Aristotle makes that show Aristotle's investment in managing a scene of reproduction that on its face could be explained as sidelining the male role.[24]

---

[23] Luce Irigaray, 'The Power of Discourse', p. 76.
[24] Irigaray, 'How to Conceive (of) a Girl', pp. 160–7.

Irigaray is concerned with the metaphysics that follow from dividing the female from the male along the lines of matter and form and associating female with irrationality and emotion and body and male with reason, mind and authority. She makes the case found in Charlotte Witt and others that Aristotle's normative metaphysics – if form and material are considered as so distinct – is both problematic and associated with gender. This reading needs to be addressed head on in order to make the case that Aristotle's biology might be a source of resisting this normative metaphysics rather than denying that there is such a problem altogether.

Mayhew makes a global dismissal of Irigaray, criticising her for making 'non-objective, sexually biased' arguments about Aristotle's text that are concerned not with 'objective conclusions about the history of gender and science', but rather 'motivated by the desire, say, to empower women'.[25] I should be clear that I do aim to empower women. I am interested in what has been called a 'retrieval' project of reading ancient philosophy to recover it from the metaphysical readings that we have inherited. The received readings of Aristotle make him the source of a binary in metaphysics that has been mapped on to gender with lasting effect. To resist the effects of that binary in metaphysics and in gender requires a return to those sources in order to suggest that they can be otherwise understood. This work is both an effort to think material in its relation to form in a way that does not oppose it to form or make it dependent on form and this work can be done at the place in Aristotle where gender is most explicitly tied to metaphysics – in the biological works.

The method of this book involves a return to arguments from the *Physics* and *Metaphysics* and *De Anima* in light of Aristotle's account of how form and matter work in generation. Instead of reading the biological texts separately from the theoretical texts, this project pursues a dialectical approach that reads the theoretical works in light of some of the difficulties raised by the biological works and takes that reading back to the biological works to explain how form and material are codependent there. This move in turn allows a reconsideration of Aristotle's metaphysics outside of a view that tries to make Aristotle consistent with Plato. I follow Aristotle's recommendation that the principles of a subject need to be adequate to the particular subject where too general principles are a problem for explaining specific problems, especially as it pertains to the meaning of Aristotle's analogies.[26] I treat Aristotle's use of analogies and metaphors in the specific context as having narrow aims rather

---

[25] Mayhew, *The Female in Aristotle's Biology*, p. 16.
[26] GA 748a8–15.

than affirming a thorough-going analogy to artifice on a global model, which is supported by the various ways Aristotle's account suggests a disanalogy between nature and artifice.

Sometimes commentators dismiss certain arguments on the basis of the motivations they find behind them. Connell argues that efforts to develop a material basis for form's work arise from a modern interest in making Aristotle amenable to contemporary accounts where inert material and movement are the source of what is. The view in this book is not that material is inert nor that movement in an efficient cause sense joined with inert material explains how generation occurs. My interest in reviving material is in no way based on a sense that modern mechanistic views of material are a solution to anything in Aristotle. Such views would only re-entrench the very problems I am concerned to address. My view is that Aristotle is unable to achieve a strict distinction in explaining the generation of animals in important ways that show us something about how Aristotle's hylomorphism works. By making this case from a dialectical reading strategy between certain passages in the theoretical works and the biological works, I argue for a conception of material that is deeply robust and interfused with the work of form.

In order to situate the problems of a binary metaphysics and gender model, I consider historical approaches to understanding gender and the ways that these approaches have been put to work in readings of Aristotle before turning to feminist criticisms of Aristotle. I conclude that the strongest criticisms are those of Aristotle's metaphysics and the possible gender implications of the normativity at work in the metaphysics. This analysis of the feminist critiques sets up the task of this book to argue for a more interdependent and cross-influencing account of matter and form in a way that does not reduce either to the other but makes them both depend on the other for their distinctiveness. The argument of the rest of this chapter will point to why that is a difficult thing for us to think.

After considering the feminist critiques, the scholarly arguments for how semen works in generation are offered. That review ends with an extended argument for the differences between Aristotle and Plato's conceptions of form. An account of form as individual and causal leads to the argument for why material in Aristotle is not the characterless stuff of prime matter. This argument from Aristotle invites an analysis of the treatments of heat and moisture in Greek myth and medicine. Such a context helps explain how Aristotle's position can be understood to show that semen is a material effort to do the work of form in a way that connects the work of heat to the work of moisture. Once Aristotle's account of form and material and his cultural context have been addressed at length, the central argument of the book is made – material has its own power and form is dependent on material to work in natural substance as exemplified in menses and semen's actions in generation. Support for this central argument is garnered in the following chapter from Aristotle's treatment of sexual differentiation and the inheritance of traits. On the basis of these claims, which join the earlier reading of form and matter in Aristotle's theoretical texts with the operations described in the biological works, the case is made for how Aristotle's τέχνη references can be understood and how his other images better reflect the interconnectedness and cross-pollination, the Möbius strip character of the relation between form and material in generation.

In her introduction, Connell writes, 'It need not be worrying for him or her to read sexist and blameworthy Aristotelian works because the content of these works can no longer hurt women . . . Obviously no one could now use the texts in this manner to actively subordinate women. Therefore, Aristotle's sexism in his biology has become politically inconsequential, which allows the feminist to feel unthreatened when analysing it.'[27] Connell is right that people should not avoid reading Aristotle or feel threatened by reading Aristotle for fear of becoming sexist. But it is not the case that Aristotle's sexism is 'politically inconsequential'. While modern science has eclipsed Aristotle's theory of generation, we continue to associate male with form and female with material and to subordinate the latter to the former. It is this division and subordination that I maintain Aristotle's own account fails to secure, which this book aims to address.

Historical Approaches to Sexual Difference

Two years after Irigaray's *Speculum of the Other Woman* was published and six years before the release of Butler's response in *Bodies that Matter*, Josine Blok took up the question of views of sexual difference in ancient Greece in her essay, 'Sexual Asymmetry: A Historical Approach'.[28] Blok argues that the way we think about sex in ancient Greece remains influenced by the ways that nineteenth-century British and German historiography addressed the 'position of women' – *querelle des femmes* – in ancient Greek literature and philosophy. In particular, the role of women in Athens posed a problem for the project of elevating the cultural status of Athens, which was central not just to nineteenth-century European classicists' research projects, but to the grander project of inscribing Europe as both the fulfilment and source of (white) civilisation. The role of women challenged

---

[27] Connell, *Aristotle on Female Animals*, p. 50.
[28] Josine Blok, 'Sexual Asymmetry: A Historiographical Essay'.

the view of Athens as the new ideal of civilisation to which nineteenth-century Britain and Germany considered themselves heirs, because if Athens is the ideal democracy and it excludes women, it is not quite the ideal. If it is not quite the ideal, then Europe does not have the unblemished inheritance it claims for itself.

Classicists responded by suggesting three possibilities: 1) Athens was not really democratic, 2) Athens did not really exclude women or 3) Athens should be judged on its own standards, not ours. None of the options were satisfactory. The second was hard to maintain. The first could not be maintained if they wanted to celebrate Athens as an ideal.[29] The last case of cultural relativism wherein scholars are invited to measure ancient Greece on its own terms makes it difficult to invoke Athens as an ideal for contemporary politics and culture.[30] The interest in what these scholars called the 'position of women' set up a dilemma no European student of the Greeks wanted, relativism on the one hand or the fall of the ideal of Athens on the other. Blok argues that nineteenth-century idealist and positivist views of difference offered the lens through which to understand sexual difference – the woman question – wherein sexual difference is transformed into sexual asymmetry and sexual asymmetry into hierarchy. The positivist optimism for a rationally orderable world viewed society and nature as exclusive, where nature and women 'as a uniform, undiversified kind' were that which could not be ordered and must be separated off in the private realm.[31] As Blok explains, whether women were isolated or not, the positivists studied them as if they were and produced the notion that they were to be so understood.[32]

While positivism separated women off as unorderable, idealism viewed sexual difference according to an essentialising binary. Men brought reason to bear on organising social life, while motherhood associated women with matter and with natural reproduction.[33] This essentialist distinction coupled with idealism's teleological history that began in ancient Greece and culminated in the European nation-state associated men with the fulfilment of civilisation, while deeming women ahistorical because they were relegated to the cycle of nature and reproduction, the 'everlasting factor of life' à la Hegel.[34] While

---

[29] For a discussion on whether democracy ever truly existed in Athens, see M. H. Hansen, *Democracy in the Age of Demosthenes*, Josiah Ober, *Mass and Elite in Democratic Athens: Rhetoric, Ideology and the Power of the People* and M. I. Finley, *Democracy Ancient and Modern*.
[30] Blok, 'Asymmetrical Sexuality', pp. 17–18.
[31] Ibid. p. 6.
[32] Ibid. p. 7.
[33] Ibid. p. 29.
[34] Ibid. p. 11.

positivism denied a true otherness making women the un-orderable, idealism located true otherness between men and women, but this difference was hierarchical, one was better than the other. As Blok argues, 'Concepts of gender and the structures that combine to define them are always asymmetrical.'[35]

When Irigaray argues that sexual difference is not true difference, she points to this asymmetry that produces the hierarchy between two poles of sex identity. In each case, the asymmetry involves a certain dependence on the subordinated pole, while at the same time deeming the superior pole superior on the basis of its capacity to transcend the subordinated pole. Butler's concern is with strategies that combat the hierarchy by maintaining the binary and showing the dependence of the superior pole on the inferior. This pole at its root is even more dependent, even more similar to that from which the reigning element of the hierarchy tries to distinguish itself. On Blok's account of the positivist and the idealist asymmetrical approaches, the reigning element of the hierarchy is always dependent on what it subordinates, culture on nature, reason on matter, men on women. Blok argues against assuming a pure binary because for the ancients, male and female are not always opposed, but the non-opposed moments fade from historical view because our lens through which we think sexual difference in the ancients has been so formed by the positivist and idealist lenses.

The effort to reconsider what form and matter mean and do in Aristotle's project in *Generation of Animals* requires rethinking the asymmetry of those categories. But not only is it a matter of rethinking the asymmetry for understanding the relation of form to matter, but also of thinking nature in Aristotle as φύσις rather than as, at its base, a τέχνη. The asymmetry of form and matter makes of form that which is imposed on material which exists solely for form, having no other being of its own.[36] The difficulty in drawing clear distinctions in how form is working and how material is working in generation points to

---

[35] Ibid. p. 40.
[36] See Helen Fielding, 'Questioning Nature: Irigaray, Heidegger and the Potentiality of Matter'. Fielding argues that Heidegger in his reading of Aristotle, which attempts to recover nature from the common view of what is given or what is opposed to convention, nonetheless 'ultimately conceives of *physis* as yet another *technē*, since in his neglect of sexual difference, women's matter is appropriated for male becoming within a being of the same', p. 2. The danger, which Fielding herself falls prey to, is to reassign nature as material in this effort to recover the feminine in a way that perpetuates the opposition between φύσις and τέχνη, feminine and masculine, as when she writes, 'For the feminine, like *physis*, is rooted in material relations, in the potentiality of our corporeal nature which [Irigaray] calls us to recognise and to perceive since perception "can never be made universal"; it is rooted in the singularity of corporeal encounters in the present', p. 21.

how Aristotle's account of nature at the microlevel of generation follows the model of φύσις and not τέχνη.

One-Sex and Two-Sex Models

Blok points to the problems that follow from defining woman as isolated from or not-man and from defining woman on her own terms. On both accounts, woman is in danger of becoming the subordinated part of an asymmetrical hierarchy. Even in Irigaray's effort to show how the one part of the hierarchy is dependent on the other, she seems to introduce the problem of positing a more true or real side of the hierarchy. These problems are echoed in the dispute over whether one-sex or two-sex models offer the best opportunities for thinking the independent identity and contribution of the feminine.

Not long after the publication of 'Sexual Asymmetries', Thomas Laqueur offered his account of the historical shifts in understanding sexual difference based on what he calls the one-sex and the two-sex models in *Making Sex: Body and Gender from the Greeks to Freud*. The one-sex model, according to which 'men and women were arrayed according to their degree of metaphysical perfection, their vital heat, along an axis whose telos was male, gave way by the late eighteenth century to a new model of radical dimorphism, of biological divergence'.[37] Laqueur argues that the performance of sex, which we tend to call gender, was what was primary in the one-sex model because the being-man or -woman on this model must be performed and maintained and could be lost, though it was the body, not social roles, that was being performed and maintained.[38] Galen presents a one-sex model that views women, not so much as a lack, but as an inversion, and less perfect version, of the male body. The difference between men and women on this model is based on place and the difference in place is due to the degree of heat. Laqueur argues that the one-sex model has no distinct female body and neither does it have a sharp boundary between the sexes, but instead a fluid boundary of degrees. The capacity associated with maleness in the one-sex model is just as much a cultural and communal capacity as a physiological capacity – the capacity to bring forth life. The male capacity is to generate life, to govern as a father would govern, giving orders and bringing order without having to engage in labour.[39]

---

[37] Thomas Laqueur, *Making Sex: Body and Gender from the Greeks to Freud*, pp. 5–6.
[38] Ibid. p. 8.
[39] Ibid. pp. 35, 54–5.

Laqueur argues that Aristotle follows the one-sex model, 'paradoxically, for someone so deeply committed to the existence of two radically different and distinct sexes', even more strictly than other ancient medical writers.[40] But he acknowledges a difference in the differences that make up this one-sex model in Aristotle, insofar as the male is the immaterial, formal and active component and the female the passive, material component. Laqueur argues that it was not the separate organs that made the difference between male and female for Aristotle, but the different roles where the male was the efficient cause and the female the material cause. For Aristotle, being male is defined by the capacity to bring forth soul, and the differences in bodies is what makes possible the difference between the efficient and material causes.[41] Laqueur argues that it does not really matter for Aristotle and the Greeks in general, who did not need, or even want, a more specific vocabulary for female anatomy, because they saw the female body as a less perfect and less powerful 'canonical body' – the male body.[42]

By contrast, the two-sex model, 'sex as we know it', according to Laqueur, was invented in the eighteenth century. On the two-sex model, the male and the female are two distinct sexes understood on their own terms. Woman, on this view, is defined according to organs she and she alone has, the uterus in particular.[43] Laqueur argues that the difference that had taken place at the level of gender and which was seen to have a fluid and dynamic effect on sex was now seen to be situated in sex and thus to have a rigid and static impact on gender. While in the one-sex model, social roles could change with changes in the body, in the two-sex model, social roles became rigidified by the affirmation of difference in the body. The biological truth of the body became the battleground for the establishment of the cultural necessity of gender roles. For Laqueur, this shift occurred not because of any new anatomical discoveries, but for historical and social and political reasons (though he does go on to discuss how the discovery of sperm and egg led to a more severe distinction between the organs that

---

[40] Ibid. p. 28.
[41] Ibid. pp. 29–30. Laqueur finds evidence for the inverted model in *History of Animals*, where Aristotle describes a tube (καυλός) that women have, like the penis but inside the body, *HA* 637a23–5. He points to the lack of precision in this description as evidence that Aristotle does not think of differences in anatomy as the basis of the difference between the sexes. Καυλός, Laqueur notes, refers to a hollow tube-like structure which in this passage could mean the cervix, the cervical canal, the vagina or the clitoris – it is not clear (Galen appears to use the same term for the penis in *On the Usefulness of the Parts of the Body*, II.324, p. 647).
[42] Laqueur, *Making Sex*, pp. 34–5.
[43] Ibid. pp. 149–50. Lacqueur writes, 'Two sexes, in other words, were invented as a new foundation for gender.'

produce those different sex cells).[44] On these terms, neither model is empirically verifiable, but rooted in particular scientific and epistemological contexts wherein the models play specific roles.[45]

The two-sex model defines women according to their own essence, but then women's essence, situated in reproduction, is devalued. Like the positivist and idealist approaches to conceptualising sexual difference in the eighteenth and nineteenth centuries, each model of difference is susceptible to asymmetry. These problems are akin to the ones Butler raises about Irigaray's reading of the history of philosophy. Irigaray argues that the version of difference that privileges male over female does indeed enable an independent definition of the female because this version involves a hidden dependence of the superior male principle on the subordinate female one. If the superior principle is dependent on the subordinate, then the subordinate principle would necessarily be something more than just not-male.[46] The form needs matter in order to manifest a natural substance, but form is conceived as better, more significant, powerful, relevant, just more, than that which is associated with the female material. The problem is, as Butler notes, by positing this outside female principle, we are in danger of making material the unformed feminine, what is most natural, the unchanging ground. When the female has her own definition, we follow a two-sex model. But the two-sex model still privileges what the male offers. Thus, the effort to assert a dependence of what appears superior on what appears subordinate seems initially to announce true difference, yet it falls back into what looks like mere lack. The subordinate principle might be truly other than the superior principle, but the worth of each remains judged in relation to the superior principle.

Even in light of these concerns, Laqueur makes an important argument about how sex in the ancient world is viewed as malleable, fluid and in need of

---

[44] Ibid. pp. 160–1.

[45] It should be noted that Laqueur's division of historical views on sex and gender has been met with stinging critique. For conflating Galen and Aristotle, see Annick Jaulin, 'Making Sex, Thomas Laqueur and Aristotle'. For errors in dating and inappropriate application to various historical epochs, see Janet Adelman, 'Making Defect Perfection: Shakespeare and the One-Sex Model' and Winfried Schleiner, 'Early Modern Controversies about the One-Sex Model'. For a book-length critique, see Helen King, *The One-Sex Body on Trial*, who argues that Laqueur selectively quotes from Galen and that Galen might be using the inversion model as a heuristic device and that Laqueur gives short shrift to the Hippocratics' positive account of women. Notably, King suggests an account of more fluidity can be found in Aristotle's *Metaphysics*, and generally that an ancient model can be found that privileges neither sex.

[46] Irigaray, *Speculum of the Other Woman*.

maintenance. Helen King echoes this fluidity with her use of the examples of Phaethousa and Agnodice.[47] Phaethousa's husband is exiled and she begins to grow hair and she stops menstruating – the absence of her husband, a change in social status, appears to threaten her femaleness. Agnodice passes herself off as a man to practice medicine, a practice that is frightening to the other male citizens for its success. As Brooke Holmes notes, these examples point to a 'gendered identity that is fluid *and* fixed, covering a continuum of traits ranging from contingent to essential'.[48]

The one-sex model shows that sex was something performed and not essential – male could slide into female *and vice versa*. If woman is defined in terms of distance from man, a fluidity exists between these positions, where the difference between them is not formal, not a difference of kind, but of degree because a difference in heat. The male has enough heat to concoct or cook the seed to the point where it can bring life into the menses. This concocting can fail or be overridden by the female and the heat that enables it can be found in both male and female bodies. In the non-hierarchical inversion account of the one-sex model, the difference between sexes is based on a range of body temperatures. While there is no true position from which difference is measured, the difference is based on degrees of heat, not some essential otherness. King writes:

> In a one-sex model, a range of body temperatures from hot to cold make possible a range of gender identities on a spectrum from the very masculine man to the highly feminine woman, a balance between hot and cold leading to a 'perfect' hermaphrodite, with spontaneous sex 'change' from female to male being theoretically possible, under the influence of increased heat: a two-sex model presents an individual with only two possibilities for their 'true sex'.[49]

While woman is devalued when defined as not-man,[50] the inversion model need not privilege the male. Further, as Holmes suggests, the one-sex model is not simply the degradation or distance of woman from the norm, but the recognition of a fundamental unity.[51]

---

[47] King, *The One-Sex Body on Trial*.
[48] Brooke Holmes, *Gender: Antiquity and Its Legacy*, p. 16.
[49] King, *The One-Sex Body on Trial*, p. 26.
[50] Esther Fischer-Homberger, 'Herr und Weib', *Krankheit Frau and andere Arbeiten zur Medizingeschichte der Frau*, cited by Lacqueur, *Making Sex*, p. 8.
[51] Holmes, *Gender: Antiquity and Its Legacy*, pp. 43-4.

Aristotle and the One-Sex and Two-Sex Models

Laqueur's book launched a debate about how to situate Aristotle between the one- and two-sex models. Aristotle does seem to define men and women according to different degrees of heat, and yet he also seems committed to a binary between form and matter, or even form and place.[52] The difficulty with categorising Aristotle as one-sex or two-sex raises questions about the differences between the models of sex as well as the difference between form and matter in Aristotle. If the difference between form and matter in Aristotle's account is based on a difference in heat, it would situate him squarely within the one-sex model. Yet Aristotle's insistence on the difference between formal and material causes points to the two-sex model. Both models are possible and insufficient ways of reading Aristotle. Both assume something that depends on understanding the difference between matter and form in the terms of the tradition where matter is wholly other than form and also lack of form.

Aristotle's account of the difference between form and matter, male and female offers us insight into how the potential for radical difference in a two-sex model transforms into asymmetry and nondifference. The apparent opposition between the one-sex and two-sex models becomes complicated in a reading of Aristotle because of the ways that male and female have been associated with form and matter and the ways that form and matter have been understood.[53] The principles of form and matter themselves imply a having and not having, if form brings something into existence and matter is defined as what is not yet formed: matter is what is for form. It chases after form because form is what material lacks.[54]

This definition of matter as what is both wholly other than and lacking form explains why, when Aristotle associates these principles with gender, he becomes associated with both the one-sex and two-sex models. Aristotle is associated with the two-sex model because he articulates two distinct principles with two distinct roles, the formal and material causes, and associates them with male and female, respectively. Yet while the two-sex model seems to portray radical difference and otherness, the hierarchy between the principles

---

[52] See Katherine Park and Robert A. Nye, 'Destiny is Anatomy'. For those who associate Aristotle with the two-sex model, see Richard Posner, *Sex and Reason*, p. 28, Holmes, *Gender: Antiquity and Its Legacy*, p. 40 and King, *The One-Sex Body on Trial*, p. 43. For those who associate Aristotle with a one-sex model against a two-sex model in the Hippocratics, see Lesley Dean-Jones, *Women's Bodies in Classical Greek Science*, p. 85.

[53] As noted by inter alia Laqueur, *Making Sex*, pp. 28–9; King, *The One-Sex Body on Trial*, pp. 40–2; and Holmes, *Gender: Antiquity and Its Legacy*, pp. 39–44.

[54] *Phys.* 192a20–4.

transforms the essence of the female as wholly other into the way she is understood on the one-sex model, not as an independent definition, but as not-male. Thus, the two-sex model falls prey to the problems of the one-sex model, when it is understood in terms of man or form as the measure.

The strategic response to the hierarchy implicit in the apparent radical difference of the two-sex model has been to challenge how fundamental that difference is by exposing the fluidity of the poles of difference. In Aristotle, the initial strict difference becomes fluid because of the way the account veers toward the one-sex model, since the male is defined as male by a certain activity that is susceptible to fail or be overcome by the female.[55] Male has a capacity that female does not, a capacity to bring life into material. The strategy of affirming this fluidity by asserting a 'sliding scale' version of difference moves Aristotle into the one-sex model where the female is a mutilated version of the male.[56] If the two-sex model is supposed to allow for true sexual difference, it does so by producing an asymmetry, making form separable from and superior to matter.[57] If the one-sex model is the solution, it results in a view of difference as simply distance from the norm. The two-sex model posits form and material as distinct principles that are different from but not reducible to the other. They are different without being contradictories, without one being that which is the negation of the other. This capacity to think difference that is not lack is part of the difficulty Aristotle confronts in thinking the relation of form to matter. The one-sex model indicates an interdependence, where one would seem to depend on the other and be capable of becoming the other, without the relation between them being strictly one of contraries because the difference is not solely of degrees but of different capacities produced by those degrees.

---

[55] See Holmes, *Gender: Antiquity and Its Legacy*, pp. 50-4.
[56] Helen King and Rebecca Flemming maintain that these are not loaded terms for Aristotle and for Galen as they are for us. King argues that 'deformed' and 'less perfect' do not carry the judgements for the Greeks that they do for us – though this is somewhat dubious given the centrality of teleology in their work, King, *The One-Sex Body on Trial: The Classical and Early Modern Evidence*, p. 41. King quotes Rebecca Flemming, who argues that these terms are their way of describing women's 'critical inability' to heat their material to the degree where it could impart life-giving breath, Flemming, *Medicine and the Making of Roman Women*, p. 119.
[57] Robert Mayhew's two-sex, two-seed model might complicate this reading of Aristotle. Mayhew argues that Aristotle sometimes uses 'seed' in a neutral way to refer to any contribution to generation, that is, to both semen and menses, so both male and female contribute seed but in different ways in *The Female in Aristotle's Biology: Reason or Rationalization*, p. 38. See Blok, 'Sexual Asymmetry'.

The two-sex model, with its essentialism, appears to establish two entirely different positions associated with form and matter in Aristotle. The difference between form and matter, which characterises the two-sex model, is, on Aristotelian terms, a formal difference. If matter is conceived as different from form in terms of differences of degrees, then the relation seems like a one-sex model. The fluidity of the difference in temperature makes the difference not one of kind, or form, but of degree, which would seem to make it a material difference. This book argues for moving beyond these models, which both result in hierarchy; such a move requires overcoming received views of matter and form in Aristotle.

The problems of the gender/sex dichotomy, the culture/nature dichotomy, and the male/female binary – as the one-sex and two-sex models illuminate – is the difficulty in finding ways to speak about material difference on its own terms. When we juxtapose formal difference to material difference we seem to be speaking of formal difference as true difference and material difference as *not as true* difference. We insist on the truth or the profundity of material difference with recourse to formal difference in a way that suggests that only formal difference could be true difference. To speak of formal difference itself as material difference seems to imply that it is not true difference, but *just* material difference, *just* a difference of degrees. The argument of this book is that material difference as both 'merely' material and profoundly true difference needs to be thought more carefully, and against all odds, Aristotle, specifically, Aristotle on sexual difference, might be a place to think this question.

In *Gender: Antiquity and Its Legacy*, Brooke Holmes makes the controversial claim that sexual difference in Aristotle is contingent, belonging to a 'realm more fluid and accidental than that of essence and principles – namely, the realm of matter'.[58] This position would see the difference between male seed and female menses as a material one. If it is material, I maintain that it reveals something about how material operates in Aristotle. Yes, it is repressed and aleatory, as Bianchi argues. Yet because this othering of material is always already characterised by the kinds of things we say characterise form, in natural things, form and material are always already interfused. Thus, Aristotle might be a site for addressing Gayle Salamon's question in her reading of Luce Irigaray's reading of Aristotle, 'What would it look like if the divide of sexual difference were not fixed in place, marked as the boundary between "male" and "female"?'[59] Such a difference entails the possibility of thinking difference within material that is not marked by form, or difference within form that is marked by material.

---

[58] Holmes, *Gender: Antiquity and Its Legacy*, p. 43.
[59] Salamon, 'Sameness, Alternity, Flesh: Luce Irigaray and the Place of Sexual Undecideability', p. 192.

Material Evidence

The difficulty of thinking the relationship of form to matter, and hence of male to female, in Aristotle is perhaps best articulated by Theodor W. Adorno, who points to the apparent contradictions in Aristotle's account of form and matter. According to a traditional reading of matter in Aristotle, 'the substrate, ὕλη, or matter, is stripped of all determinateness, so that it becomes something entirely empty, and comes extraordinarily close to the non-existent in Plato'. Form makes matter determinate and knowable, yet form is dependent, 'form is always the form of *something*'.[60] Both material and form have a problem with relation to the determinate. Matter seems indeterminate without form, but form is not-yet-materialised without matter. While form might be thinkable on its own, it depends on material in natural substance. Material can exist without form in its basic existence as elements and it is able to resist form, which points to a particular power that belongs to matter. While Adorno argues that Aristotle fails to think the mediateness of form, it seems that Aristotle's account leads to these impasses because he is working through this mediateness in his metaphysics. Against a Platonic conception of form as that which exists prior to material in natural substances, for Aristotle form requires material in order for the natural substance to come to be. This requirement is the dependence with which Irigaray is concerned that makes form less independent of matter than it would seem to need to be to be understood as conceptually distinct, as form is supposed to be understood. Thus, it would seem that both the epistemology of form and the ontology of form depend on a relation to matter that would challenge our previous understanding of both form and matter.

In *Aristotle on Substance: The Paradox of Unity*, Mary Louise Gill similarly approaches the project of rethinking the robust contribution of material in Aristotle's metaphysics. Much as Adorno recognises material's capacity to resist form, Gill argues that matter is a serious threat to the conceptual and ontological unity and hence to the substantiality of material substances of which they are a part.[61] Gill points to how the question of material's role in

---

[60] Theodor W. Adorno, *Metaphysics: Concepts and Problems*, pp. 61–2. While this seems to have been the general view when Adorno published this book in 1965, the idea that matter has its own being and character can be found in Heidegger dating back to the 1920s. Heidegger reads ὕλη in Aristotle in a positive co-determining sense, describing its role in the work of a natural being, emphasising that it is not not-being or indefinite stuff or the not-being or lack of form in *Basic Concepts of Aristotelian Philosophy*, pp. 140–56. But Heidegger's account still describes ὕλη as dependent and reliant on εἶδος to show itself. Heidegger says he is not treating ὕλη as stuff but then speaks of what natural beings consist of as the elements which he then describes as stuff, p. 154.

[61] Mary Louise Gill, *Aristotle on Substance: The Paradox of Unity*.

substance joins questions of change in Aristotle to questions of the nature of substance. The concern is that answers to the first question controvert the answers to the second question, producing what Gill calls the paradox of unity. This paradox shows that Aristotle's corpus is torn between ontological and conceptual priority, which Aristotle maintains all natural substances have. Material poses a challenge to the conceptual priority because the natural substance is a composite, so one part of the composite seems primary rather than the natural substance itself. Aristotle conceives of natural substance as a composite in order to explain how substance can be generated, an explanation that requires a reference to material that persists through change while the form that previously was not in certain material comes to be in that material. Material explains how there can be what Gill calls horizontal unity in Aristotelian substance, a continuing relation to what was before, rather than coming to be from nothing. Form explains how at any given moment a composite is a vertical unity, a substance that is a primary entity. Aristotle's account of change requires that there be horizontal unity; his account of substance requires that there be vertical unity. Both accounts involve a certain conception of material – one that remains in some sense through change because it is capable of taking on new form and one that has some sense of its own character such that the unifying work of form is necessary. The problem is that the conception of matter that explains how substance generates seems at odds with the conception that explains how substance is what it is. The unity through time seems to compromise natural substance's unity at any particular moment. Material is needed to do distinct work from form but their distinct roles would seem to make natural substance conceptually plural rather than unified in one account. Gill argues that Aristotle resolves this dilemma through an argument for how Aristotle explains the persistence of matter in substantial change through to substantial being.

A similar problem is at work in understanding how form in animal generation comes to have formal power through material processes. Form does the unifying work of individual substance, but this unity is compromised by the necessary work of material in enabling the generation of substance. Material seems to similarly threaten the vertical unity of substance if it is to be explained through the distinctive power of form. In generation, a certain kind of material comes to do the work of form, which becomes form by virtue of having a different capacity than the material that it was before being worked up into form, or the material that it will work on in the mother which has its capacity by virtue of not being so worked up into form. Form's capacities work through and in material in semen, and yet as form, must be distinct from it. Yet if form does not work from within the material processes, then it is not clear how form comes to work on material. But if form does work from within material processes, how it is a different principle from material is not evident. The questions of how form can become form through a material process follow from understanding the formal principle as not unified with material. But if form is not becoming form from material processes, it is not clear how it comes to do its work in a way that makes it unified with material substance. The perduring of material through generation explained how what was not comes to be. The material dependence of form on matter explains how what was contributes to what comes to be. But this dependence raises the question of how the form that works on the newly formed substance works as an independent form.

In this book, I will make a case for how to understand these passages to show that the difference within material becomes the source for a difference between what acts as form and what acts as material in a way that makes it difficult to distinguish between the strictly material and the strictly formal. This book invokes Elizabeth Grosz's image of the Möbius strip where what appeared to be opposed finds itself on the same side to think the relation of form and material in Aristotle's biology. Grosz discusses this relationship in terms of the mind and the body in *Volatile Bodies*:

> The Möbius strip model has the advantage of showing that there can be a relation between two 'things' – mind and body – which presumes neither their identity nor their radical disjunction, a model which shows that while there are disparate 'things' being related, they have the capacity to twist one into the other. This enables the mind/body relation to avoid the impasses of reductionism, of a narrow causal relation or the retention of the binary divide.[62]

On the Möbius strip account, form and matter are neither identical nor opposed in a binary that leads to asymmetry. Material has significance of its own in a way that connects it to form without reducing or opposing it to form. Historically philosophers have had great difficulty in speaking of material as significant without reference to form. On the one hand, it seems that to say that matter matters because it is always enformed would make matter material because of its relation to form, which makes it significant only through form, which has the real signifying power, rather than material. On the other hand, to say that material is significant as a base construes and

---

[62] Elizabeth Grosz, *Volatile Bodies: Toward a Corporeal Feminism*, pp. 209–10.

thereby invents material as a ground, which is associated with the role of the feminine, as Butler argues is the case for Irigaray. Against these positions, this book stakes out a third position wherein material's significance is rooted in specifically material power and its difference from form in generation is a material one in the sense of a difference of degrees and a formal one only if the difference between form and matter is considered a formal difference. It is in holding these material and formal differences between matter and form together that Aristotle's account points to the model of the Möbius strip.

1
Feminist Critics of Aristotle's Biology and Metaphysics

The Feminist Disputes

In this chapter, I consider the arguments among feminists and readers of feminist critics over how to understand Aristotle's biological texts. The last quarter of the twentieth century witnessed a surge of interest in Aristotle's biology. Many commentators during this period argue that Aristotle's biology was outright misogynist. One argument for this misogyny is the inequality of the contributions of male and female in generation. This view is contested by others who argue that Aristotle offers more of a role to female animals than his predecessors and contemporaries do. Commentators disagree about the extent to which Aristotle's views of male and female in the biology affect the rest of his corpus. Some find that Aristotle's account of their differences in the biology leads him to arguments about the irrationality of women in the ethical and political works, while others claim that Aristotle never offers arguments for those views and that even if his writing could be construed as an argument in defense of these claims, that they are not based on the arguments of the biology. Many commentators raise the question of whether Aristotle's metaphysical framework requires two sexes. Some commentators argue against the view that Aristotle associates male with form and female with material altogether. Other defenders of Aristotle find feminist critics' readings unpersuasive because they depend on asking questions of Aristotle that he would not ask himself. One response to that charge is that regardless of whether Aristotle would have asked these questions, his metaphysics that informs and is informed by his account of gender has normative implications that can be examined on contemporary ethical terms. Defenders of Aristotle's metaphysics argue that the reduction of material to an insignificant contribution is based on poor understandings of material in Aristotle. They similarly dispute the view that artifice is the best way to understand the role of form and matter in generation. This chapter articulates the criticisms and responses in order to present the landscape of the critical conversation around generation in Aristotle's biology.

Feminist Critics Who Say Aristotle's Biology is Sexist

It has become almost commonplace to consider Aristotle's biology misogynist. By 1989, Eva Browning Cole can write in a review of Page DuBois's *Sowing the Body* acknowledging, 'We have been accustomed to regard Aristotle as the fountainhead of one long tradition of western misogyny.'[1] It seems sufficient for Mary Mahowald to make this case by simply listing passages from *Generation of Animals*.[2] Martha Nussbaum calls Aristotle's biology 'both misogynist and silly'.[3] Eva Keul calls Aristotle 'one of the fiercest misogynists of all times, obsessed with the need to prove that women played no genetic part in reproduction'.[4] And indeed, it is the case that historically, Aristotle's biology has been used to defend subordinate social, cultural and political positions for women.[5]

Both those appealing to Aristotle to defend the inequality of the sexes and feminist critics have assigned views to Aristotle that are not always borne out in the text. One common argument is that Aristotle finds the female contribution to be merely the container in which generation and growth occur.[6] This position leads some to ascribe to Aristotle the preformationist account of the Hippocratics that he explicitly rejects.[7] For those willing to

---

[1] Eva Browning Cole, 'Review of Page DuBois, *Sowing the Body: Psychoanalysis and Ancient Representations of Women*', pp. 88–9.
[2] Mary Mahowald, *The Philosophy of Women*, pp. 266–72.
[3] Martha Craven Nussbaum, 'Aristotle, Feminism, and Needs for Functioning'.
[4] Eva Keul, *The Reign of the Phallus*, p. 405.
[5] For a litany of such uses, see Maryanne Cline Horowitz, 'Aristotle and Woman'.
[6] Ruth Bleier argues that centuries of scientists were misled by 'the 20-centuries-long concept, stemming from Aristotle, that women, as totally passive beings, contribute nothing but an incubator-womb to the developing fetus that springs full-blown, so to speak, from the head of the sperm', p. 3. See also Caroline Whitbeck, 'Theories of Sex Difference', p. 55. Horowitz, 'Aristotle and Woman', p. 192; Velvet L. Yates, 'Biology is Destiny'; Keul, *The Reign of the Phallus*, p. 145; and Page DuBois, *Sowing the Body*, p. 126.
[7] Jean Bethke Elshtain describes the male's work in Aristotle's account as one in which the male 'implants the human form' in the female and one in which the male 'deposits within the female a tiny homunculus for which the female serves as a vessel until this creature matures. The female herself provides nothing essential or determinative', in *Public Man, Private Woman: Women in Social and Political Thought*, p. 44.

acknowledge some female contribution, it is limited to a passive material.[8] Another common view among feminist critics is that Aristotle believes that only male has seed,[9] and that this lack of seed is due to the female's relative coldness.[10] One consequence of the view that the male alone contributes seed is the view that Aristotle thinks reproduction is based on the role of the father alone and that it aims toward reproducing the male, a view that is pretty widely held.[11]

Some feminist commentators argue that Aristotle's sexism in the biology influences his whole corpus, especially his account of women's subordinate virtue and political roles in the practical works.[12] Others take Aristotle to offer a justification for widely held Greek social attitudes.[13] A host of feminist commentators argue that Aristotle associates the male with mind and reason by

---

[8] Sue Blundell, *Women in Ancient Greece*, p. 106. Andrew Coles argues that since pulsation and pneumatisation are the source of the heat in the blood, the male could not be hotter, so this process explains vital heat as the source of life but belies Aristotle's sexism in arguing that the female's only contribution is matter, which Coles argues is Aristotle's 'stage-managing' of his hylomorphism mapped onto Aristotle's gender division, since the account of pulsation in the blood shows that the female contributes vital heat as well, Andrew Coles, 'Biomedical Models of Reproduction in the Fifth Century BC and Aristotle's *Generation of Animals*', p. 88.

[9] Helen King, *The One-Sex Body on Trial*, p. 40. Lesley Dean-Jones writes, 'It is not sufficient to cite Aristotle's sexism here. The Hippocratic doctors were equally convinced of female inferiority. Their theory of female seed did not in any way amount to saying that women were equal to men (note their description of female seed as weak seed); it was merely the most obvious way of explaining the undeniable resemblance many children bore to their mothers. The possibility of resemblance to the mother was just as obvious to Aristotle, but he appreciated the difficulties inherent in the unsophisticated theory of equivalent male and female seed', in *Women's Bodies in Classical Greek Science*, p. 179.

[10] Prudence Allen, *The Concept of Woman: The Aristotelian Revolution 750 BCE–1250 AD*, p. 34.

[11] Commentators who maintain males are the complete form of the species and most substances and females are incomplete forms and so only substances in a qualified sense inter alia Errol G. Katayama, *Aristotle on Artifacts: A Metaphysical Puzzle*, p. 253; Gad Freudenthal, *Aristotle's Theory of Material Substance*; Gill, *Aristotle on Substance*, p. 33; Pierre Pellegrin, 'Aristotle: Zoology Without Species'.

[12] Lynda Lange, 'Woman is Not a Rational Animal: On Aristotle's Biology of Reproduction', in *Discovering Reality*; Susan Moller Okin, *Women in Western Political Thought*, p. 80; Yates, 'Biology is Destiny'.

[13] G. E. R. Lloyd, *Science, Folklore, and Ideology*, p. 4. Coles seems to agree in 'Biomedical Models of Reproduction in the Fifth Century BC and Aristotle's *Generation of Animals*'. Dean-Jones, *Women's Bodies in Classical Greek Science*, Sophia M. Connell, *Aristotle on Female Animals* and Daryl McGowan Tress, 'The Metaphysical Science of Aristotle's *Generation of Animals*' argue that Aristotle is rejecting Hippocratic claims.

means of the male's association to form and the female with emotion and the body by means of the female's association with material.[14]

Other feminists argue that Aristotle is sexist, but his whole philosophy is not. Rhoda Kotzin finds Aristotle's accounts of women to be inconsistent with other elements of his philosophy.[15]

Critics Who Say Aristotle's Biology is Not Sexist

Feminist critics argue that the female is simply a container for growth, only the male contributes seed, the female does not contribute anything that contributes toward the life and form of the offspring, what the female contributes makes the female inferior, and the differences between the male and female point to an asymmetric contrary relationship between them.

Defenders of Aristotle have responded to each of these claims. Many commentators argue that Aristotle is pretty straightforward about the female contribution as material, a position that would challenge the claim that Aristotle is merely trying to justify contemporary Athenian gender roles, since his account gives more of a role to the female in generation than the Hippocratics.[16] In fact, Aristotle's most explicit statement of the difference between the male and the female roles is, 'By a "male" animal we mean one which generates in another, by "female" one which generates in itself.'[17] This is the statement that feminist critics have taken to mean the female is only a container. But this definition is not limiting the female to the container, but rather distinguishing the male who generates in another from the female who generates in herself. Tress argues that Aristotle elevates the female by arguing in *Generation of Animals* I that there are two principles of generation – male and female.[18]

Supporting this mutual contribution, a host of commentators argue that Aristotle considers the female contribution seed.[19] This view is often obfuscated

---

[14] Inter alia Donna Wilshire, 'The Uses of Myth, Image, and the Female Body in Re-visioning Knowledge'; Donna Haraway, 'Situated Knowledges: The Science Question in Feminism and the Privilege of Partial Perspective'; Yates, 'Biology is Destiny'.

[15] Rhoda Kotzin, 'Aristotle's Views on Women'.

[16] Johannes Morsink, 'Was Aristotle's Biology Sexist?' and Tress, 'The Metaphysical Science of Aristotle's *Generation of Animals*', p. 309.

[17] GA 716a13-14. See Tress, 'The Metaphysical Science of Aristotle's *Generation of Animals*', p. 314 and L. Aryeh Kosman, 'Male and Female in Aristotle's *Generation of Animals*'.

[18] Tress, 'The Metaphysical Science of Aristotle's *Generation of Animals*', p. 316.

[19] Anthony Preus, 'Science and Philosophy in Aristotle's *Generation of Animals*', pp. 8–9. See also Marguerite Deslauriers, 'Sex and Essence in Aristotle's *Metaphysics* and Biology', p. 148; Robert Mayhew, *The Female in Aristotle's Biology: Reason or Rationalization*, pp. 34–40; Tress, 'The Metaphysical Science of Aristotle's *Generation of Animals*', p. 323. See GA 724a17-20, a35-b6.

by the translation of σπέρμα as sperm and not seed, a translation that leads Tress to argue that 'Sexism here has its sources, not in Aristotle, but in his nineteenth and early twentieth-century translators.'[20] Aristotle calls the menses analogous to semen as seed and only seems to deny that the female contributes seed when referring to fully concocted seed. At 749b3-9, seed refers to both semen and menses.[21] At 750b4-5, Aristotle says that spermatic matter is present in the female and he describes both male and female contributions as residues in contradistinction from waste-products in *GA* I.18. While Aristotle does say at *GA* 724a7-10 that the female does not emit seed but is the cause of generation in some other way, one interpretation is that the female contributes seed but does not emit it, referring to the Hippocratic view that both emit seed and that the seed mingles to produce the embryo.[22] This view – that both emit the seed, sending it out of the body – Aristotle rejects, but not the view that the female contributes something to generation.

Marguerite Deslauriers argues against those critics who claim that Aristotle justifies the political subordination of women to men by their biological differences, specifically because Aristotle does not depict sexual difference as difference in form or species.[23] She advocates a view of sexual difference that is nonaccidental and yet nonessential, since the differences are peculiar to the genus animal, but are located in matter and not in form. On Deslauriers' reading, the contrary of sexual difference characterises male and female *principles* (which is what Aristotle references when he says that male and female differ in λόγος), not *animals*, as opposites. Sexual difference belongs to the genus animal rather than the individual species which makes it a material contrary belonging καθ' αὑτό, nonaccidentally, to the material.[24]

---

[20] Tress, 'The Metaphysical Science of Aristotle's *Generation of Animals*', p. 326n30.
[21] See *GA* 766b12-15.
[22] Mayhew, *The Female in Aristotle's Biology*, p. 34; Tress, 'The Metaphysical Science of Aristotle's *Generation of Animals*', p. 325. *GA* 728b21-5. Cf. Connell, *Aristotle on Female Animals*, pp. 95-120.
[23] Deslauriers argues that Aristotle does not argue that women are essentially inferior, nor does he make an argument for the deficiency of the female in 'Sex and Essence in Aristotle's *Metaphysics* and Biology'. See Devin M. Henry, 'How Sexist is Aristotle's Developmental Biology? *Phronesis* 52 (2007), pp. 251-69; Sophia M. Connell, *Aristotle on Female Animals*, p. 25 and Connell, *Aristotle on Female Animals*, p. 33; Tress who argues from *Metaphysics* X.9 that it cannot be said that male and female differ substantially, 'The Metaphysical Science of Aristotle's *Generation of Animals*', p. 315. Against Spelman, 'Sex and Essence in Aristotle's *Metaphysics* and Biology', p. 155.
[24] Deslauriers, 'Sex and Essence in Aristotle's *Metaphysics* and Biology', pp. 145-6, 152. Cf. Connell, *Aristotle on Female Animals*, pp. 170-2. See *Metaphysics* 1058b23-4, 1078a5-7.

Within the biology, Deslauriers concludes that the difference between male and female is a difference of degree based on the capacity to concoct. Both the male and female concoct food into blood. The male concocts the blood into semen through a kind of natural heat, made possible by a special kind of air in the heart and testes. Since the heart works in the same way in male and female, the difference that matters for sexual difference is made in the genitals – the male organ is able to produce the friction to produce the heat and females have organs that do not produce that heat. The male is the moving cause only because he can better heat the male contribution. An animal becomes male or female when the sex organ is formed, but since the organs are accidental and the capacity is acquired with the organ, the difference in capacity is accidental. If the differences were not accidental, they would be transmitted with the form, which does not occur. The determination of sex depends on whether the male seed prevails over the female seed but, Deslauriers argues, the contributions themselves are not sexed. For these reasons, Aristotle cannot maintain that sex is essential. Deslauriers rejects the view that Aristotle's account of sex determination indicates that the male contribution is male and the female contribution female before conception because this view would entail that the male contribution be female when a female offspring results. Instead, she argues, the movement of the male residue is potentially male or female in every case, which supports Aristotle's case against preformationism.[25]

Following Deslauriers, who offers a material account of the difference between the male and female in terms of degrees of concoction, a number of commentators point to the material differences between male and female based in vital heat. Andrew Coles argues that for Aristotle the male body is not hotter than the female body, rather the movement of blood through the testes in the male makes the semen hotter than the menses.[26] On this account, the movement in male and female is the same, since both are caused by the pneumatisation in the blood.[27] This view leads Coles to conclude that the difference between male and female contributions is not a difference between form and matter.

Devin M. Henry distinguishes the male and female in terms of their capacity or incapacity to produce semen (γονή), which is made possible by natural heat, which is found in both men and women but hotter in the hearts of men. Henry argues that the male and female both have the ability to produce σπέρμα by

---

[25] Deslauriers, 'Sex and Essence in Aristotle's *Metaphysics* and Biology', pp. 147-51.
[26] Coles, 'Biomedical Models of Reproduction in the Fifth Century BC and Aristotle's *Generation of Animals*', pp. 66-7.
[27] Against Cooper's view that there is a difference between movement in the matter and movement in the blood in John M. Cooper, 'Metaphysics in Aristotle's Embryology'.

concocting the surplus of blood in the heart, but differing abilities to concoct it, which is why the menstrual fluid is colder, more fluid and greater in volume. Henry concedes that this ability makes the female inferior since she does not have the capacity of the rest of the species, but he denies that this difference makes the female a failure. Sex determination, Henry argues, is a result of non-teleological necessity because sex is not part of the substance of an animal.[28]

A host of commentators conclude that Aristotle does not need to explain that there are two sexes as he does to be consistent with his metaphysical framework, but explaining things as he does shows that he is falling back on sexist views or ideology.[29] He maintains that the efficient or moving cause is superior to the material cause, and by associating these causes with the sexes, maintains that the male is superior and so better separate from the inferior female, but the first couplet need not be mapped onto the second.

L. Aryeh Kosman argues against Henry's claim that the male contains the form that the male parent in Aristotle's generation cannot be said to contain the specific form of the adult animal in any sense in which the female menses does not. The sperm is the first mover, not the source of form in the animal. Kosman describes a material account of the work of the semen as beginning a process in which the menses is an equal partner.[30] On Kosman's account, both the male and the female contain the parts of the offspring potentially, and neither contain them in actuality. The male contribution produces movement which works on the female contribution. The female part is not incapable of producing offspring, it is incapable of starting the process of generation. Kosman concludes any asymmetry of power is limited to beginning the process, not to the process of generation as a whole, in which the female is one of the two principles.

Some scholars point to the account of the female as that which generates in itself, and argue that matter is derivative of that role.[31] According to this view,

---

[28] Henry explains Aristotle's view of woman as monster and deformity with reference to other places where Aristotle sees other animals – like seals and puppies – as falling short of a certain sense of completion. Henry, 'How Sexist is Aristotle's Developmental Biology?' pp. 255–65.

[29] Dean-Jones, *Women's Bodies in Classical Greek Science*, pp. 187–8; Coles, 'Biomedical Models of Reproduction', p. 84; Henry, 'How Sexist is Aristotle's Developmental Biology?' pp. 264–6; Witt, 'Form, Normativity and Gender in Aristotle', p. 129 each argue that it is Aristotle's commitment to maintaining a certain gender social structure that leads him to this view, while Mayhew argues that there are non-gendered reasons for him to make this argument to claim that it is not an ideological position.

[30] Kosman, 'Male and Female in Aristotle's *Generation of Animals*', pp. 255–6.

[31] Ibid. p. 256 and Tress, 'The Metaphysical Science of Aristotle's *Generation of Animals*', pp. 314–15.

Aristotle deduces that the woman contributes material because he observes that the offspring is generating in the female and material in the form of nourishment is required for growth.[32]

Tress argues that feminists go wrong by asking inappropriate questions of Aristotle's biology, like who controls reproduction, a question whose aim is political, neglecting the metaphysics which, in her view, are not concerned with issues of power. Tress argues feminist critics read Aristotle out of context, suggest a standard of activity to measure equal value, and underappreciate the role of material in Aristotle's metaphysics by assuming a post-seventeenth-century sense of matter as inert.[33] These critics suppose that the female for Aristotle is not capable of generating anything because she is not capable of making anything else come alive by itself, a standard that would make nothing capable of generation in Aristotle since nothing comes alive by itself.[34] Against what she considers the egregious and most widespread view that Aristotle argues that 'in the act of generation the passive, potential female matter is bestowed with life by the actualised and soul-bearing form of the male', Tress argues that Aristotle describes the female and male fluids both as potentialities, and the offspring that results as an actuality.[35]

Tress argues that the material contribution of the female in generation is not raw but 'specialized and highly refined' because it potentially contains all parts of the body.[36] The problem with the pangenesis and preformationist views of Aristotle's predecessors, and, according to Tress, any materialist view, is that they fail to account for the principle that organises them.[37] Tress explains that the preformationist view was used in Greek society to deny a role in generation to women, leading to the view that woman is merely a container. While Empedocles's pangenesis might seem to offer a position to both male and female, it still fails to offer a unifying principle to the parts

---

[32] Mayhew, *The Female in Aristotle's Biology*, pp. 39–40.
[33] Tress, 'The Metaphysical Science of Aristotle's *Generation of Animals*', pp. 309, 321, 321n25. On the view that feminist critics assign to Aristotle a notion of matter as inert see also Connell, *Aristotle on Female Animals*, pp. 124–5.
[34] Against Cooper, 'Metaphysics in Aristotle's Biology', p. 17.
[35] Tress, 'The Metaphysical Science of Aristotle's Generation of Animals', p. 332; see Connell on offspring as actuality, *Aristotle on Female Animals*, p. 172.
[36] Tress, 'The Metaphysical Science of Aristotle's *Generation of Animals*', p. 327; see GA 737a23.
[37] Tress, 'The Metaphysical Science of Aristotle's *Generation of Animals*', p. 318, 'Aristotle Against the Hippocratics on Sexual Generation: A Reply to Coles'.

drawn from all parts of both contributing bodies. For Tress, the body aims at the soul from the beginning, which seems to imply that the soul – the unifying principle – in some sense pre-exists the body that aims toward it. If both contributing factors are the organising principle, then they would need yet another organising principle to bring them together. If both contributing factors are matter, they need an organising principle to bring them together. One principle must be material and the other what organises it. The question is, how does that principle work? Tress takes issue with accounts that try to drop out the metaphysical causes and make form merely mechanical movement that produces capacities in the semen, a position that Tress argues Aristotle regularly rebuts with recourse to teleological explanations.[38]

Tress argues that the male's contribution works not through its material but through 'some disposition, and some principle of movement'.[39] The male's contribution, Tress argues, is immaterial, working as a tool that imparts motion.[40] Tress argues that, indeed, Aristotle maintains that the sexes are distinct because it is better for the higher to be separated from the lower – she offers a metaphysical account that supposes that all of the natural world aims toward what is eternal and divine, and is acted upon by the divine and beautiful to the extent that it can be affected, in such a way that leads natural things to be better, ensouled and alive to the extent that they can be.[41] So form is separate from material to allow it to be superior, a view that she realises follows from Aristotle associating male with form and privileging it over the female, arguing,

> In this instance, as in some others in *GA*, we have evidence of the infusion of Aristotle's commonplace views into his more careful philosophical thinking. There is, after all, no reason he has given to hold that those whose generative fluid transmits form are superior as a kind or class to those whose generative fluid transmits matter. It is true, of course, that within Aristotle's metaphysics form has preeminence over matter, but neither Aristotle nor his reader is entitled, on the basis of what is established so far in *GA*, to conclude that this preeminence accrues to male creatures *qua* male. Recall from book 1 that organisms are male

---

[38] Tress, 'Aristotle against the Hippocratics on Sexual Generation', pp. 322–9.
[39] *GA* 726b20–1.
[40] *GA* 730b15–19.
[41] *GA* 731b25–8, *Meta.* 1072b3–4.

and female not as a whole body but only with respect to their generative roles; he establishes further that both male and female are fully members of the same species, one implication of which is that both male and female individuals possess the same 'substantial formula' of matter and form ... There is no philosophical justification, then, for this remark about the superiority of the male.[42]

Tress thus reads Aristotle to draw conclusions according to his cultural norms that are not supported by his work. She concludes that such claims can be sidelined in favour of the ones that are supported.

Tress argues that Aristotle addresses the separation of the sexes or the need for two sexes in order to explain how the seed forms the offspring, which poses a problem because something external (immaterial) is forming life which it would seem difficult for it to do without engaging the material materially. On the one hand, the form could already be in the material, which would lead to something like pangenesis or preformationist views, or on the other hand, it is external and unable to work on what needs forming. Tress maintains that for Aristotle the semen does not contribute material to the offspring. Instead of articulating the problem of how form moves into material in terms of an internal/external dichotomy, Aristotle explains the work of form and matter in the fetation in terms of actuality and potentiality. None of the physical properties on their own can explain the λόγος or the highly organised nature of the organism, and she argues that the λόγος derives from the parent who imparts movement and is in actuality what the offspring is in potentiality. Tress's explanation is that the male and female contributions and the fetation possess the nutritive soul potentially, but only the semen possesses the sentient soul. Reason enters from outside because the part of the soul that is reason is not material, but still the soul needs some physical substance which is more divine than the elements, some element that is 'highly rarefied and unique', like the αἰθήρ or aether in the stars.[43] This substance, Tress argues, is the element of which the stars are made and it has nothing to do with fire. Tress explains that the way that heat links between the body and soul is not explained because, 'Nature itself is a coherent whole; no fundamental division exists in nature which would be in need of philosophical repair.'[44] If the male contribution is material only in a sense that is connected to the celestial, the female contribution is prime matter in a cosmic sense, 'the most primordian stratum of the

---

[42] Tress, 'The Metaphysical Science of Aristotle's *Generation of Animals*', p. 331.
[43] Ibid. pp. 331-5.
[44] Ibid. p. 335.

natural world'. Out of this axis of the male αἰθήρ and the female primordian stratum, the intersection of Ouranos and Gaia seems to meet in generation:

> The primordian, preelemental power that inheres in prime matter expresses itself (as potentiality) in the adult female's generative material. The active, moving power that is unique to the heavens expresses itself (as potentiality) in the male generative material.[45]

Feminist Critics and Defenders of Aristotle's Metaphysics

Charlotte Witt argues that the problem is with Aristotle's metaphysics: his hylomorphism is normative, and 'Aristotle attaches the gender norms of his culture to hylomorphism.' As Witt explains, these views are normative because for Aristotle, nature is normative, and any account of nature that fully reflects the reality of the natural world must be a normative theory. Witt argues that both Aristotle's account of form conceived as function and his framework of teleology that distinguishes between what is potential and what is actual constitute the normativity of Aristotle's conception of nature. She argues that Aristotle's account of form in terms of function and his use of teleology specifically in the concepts of potentiality and actuality 'readily lend themselves to more explicit and concrete political coloring (e.g., to gender associations)'.[46]

Witt's position is not that of what she calls the 'bizarre view', perhaps attributable to Irigaray, that Aristotle saw reality or nature as gendered. This position, which she calls the 'gendered interpretation of hylomorphism', supposes that since form comes from male form is male and since matter comes from female matter is female. Similarly, Witt rejects Okin's view that women exist only for the good of another. 'On the contrary, for all natural substances – men, women, dogs and cats – the life functions whose principle is soul are for their own good.' Witt argues that Aristotle's concept of functional form, at least, does not support the view that women are functional for men, a view that is metaphysically inconsistent or incoherent for Aristotle since all natural substances are constituted of both form and matter, so if matter were female and form male, animals would be hermaphroditic and human beings would differ in form since the difference between matter and form is formal – while belonging to the same species. Witt rejects this view in part because she does not think the political

---

[45] Ibid. p. 338.
[46] Charlotte Witt, 'Form, Normativity, and Gender in Aristotle: A Feminist Perspective', p. 121.

and gender implications of Aristotle's metaphysics would be resolved if the formal and material principles were not associated respectively with the male and female.[47]

Witt argues instead that form itself by being inherently normative institutes a division between those kinds of things that are more formed and those kinds of things that are less formed. Things are defined by their functions, the form is the principle of the function, what causes it and makes it possible. The form is what actualises the living being as what it is and makes it capable of being what it is. Form gets us to the end, matter pursues the form. Form is a good as an end, as the function that shows what a thing should do, and it is better than matter. For Witt, Aristotle is developing the political and ethical implications of his metaphysical account when he says that women are defectively rational, not having their forms fully formed. That the form is associated with male and the matter with female depends, Witt argues, on the way that men and women are unequally valued in Aristotle's culture, but even so, redressing that wrong does not redress a metaphysics that divides between what is formed and what is incompletely formed. Witt argues that while Aristotle's normative valuation of form does lead him to divide between a better and more divine principle and an inferior material principle, this need not necessarily have been written onto gender difference, though it was, and merely distancing it from that difference does not change the fundamental hierarchical structure embedded in the account Aristotle offers.

Other scholars agree with Witt that the metaphysics are the problem. Judith M. Green argues that the metaphysics are based on a verticality that privileges the sides of contraries that most achieve the organism's function, finding the hierarchy and normativity inherent in Aristotle's teleology.[48] In this schema, participation in the feminine principles is the cause of all downness (moving away from the vertical upness achieved in substance being). While Cynthia A. Freeland articulates an interdependence of matter and form, where matter yearns for form, and then form is matter for the next level up, she argues that Aristotle still yearns for a position of mastery articulated in pure actuality. Freeland describes pure potentiality and pure actuality as Aristotelian fantasies, 'Aristotle longs to be a similar being [to the male

---

[47] Ibid. pp. 124–32.

[48] Judith M. Green, 'Aristotle on Necessary Verticality, Body Heat, and Gendered Proper Places in the Polis: A Feminist Critique', pp. 71–2, 82. This normative view of Aristotle's metaphysics is held even by scholars less critical of Aristotle, like Abraham P. Bos, who describes material for Aristotle, not as a negation of Being, 'but as typified by deficiency of being and desire for being', *Aristotle on God's Life-Generating Power*, p. 37.

principle] who is self-creating, self-fulfilling, self-absorbed, self-delighting – a holy God unnourished and unborn of woman. This is an extraordinary, and in some way ultimate, construct of the male metaphysicians' mind.'[49]

Claire Colebrook argues that Aristotle's teleology produces the concept of the 'proper potentiality' in her analysis of Irigaray's reading of Aristotle, writing: 'The potential to reason defines what we ought to become in order to actualise our being; and this potential to reason means that we only live our lives fully when we move beyond mere nutritive and perceptive movements, and order our lives as a whole.' Colebrook argues that a life of meaning is one in which the self forms itself as a coherent and ongoing narrative whole. Wholeness, and completeness and togetherness are privileged as substance, form, being. Colebrook finds in Irigaray's reading of the Greeks 'the possibility of a life whose relations have no proper potential, a life dispossessed of truth and presence, a life whose relations will have to be effected and achieved'.[50] Colebrook captures the sense in which Irigaray is concerned that proper potentiality, reason, telos and actuality all require something that underwrites them for them to be possible, while this underwriting is also suppressing what grounds.

While Connell argues that Irigaray's reading is impressionistic,[51] Bianchi argues that Irigaray's analysis of Aristotle's metaphysics is borne out in Aristotle's biological corpus. Bianchi argues against those who say the metaphysics are normative and hierarchical but not necessarily gendered, saying of the connection of Aristotle's metaphysics to gender hierarchy: 'It is utterly systemic and constitutive.'[52] Her view departs from the notion that Aristotle's conception of nature is fundamentally one of τέχνη, where form imposes itself from the outside on matter. Bianchi argues that while φύσις seems to internalise its source, in generation,

> this source of internal change is reexteriorized and literalized as the father, thus reestablishing an appropriative technical relation at the heart of *phusis*, insofar as the male is defined in *Generation of Animals* as that which generates in another, while the female generates within herself, providing only the site and the material substrate for the offspring.[53]

---

[49] Cynthia A. Freeland, 'Nourishing Speculation: A Feminist Reading of Aristotelian Science', p. 177.
[50] Claire Colebrook, 'Dynamic Potentiality: The Body that Stands Alone', pp. 178, 179, see *Meta.* 1047a4–7, *EN* 1142b31–3.
[51] Connell, *Aristotle on Female Animals*, p. 19.
[52] Bianchi, *The Feminine Symptom*, p. 27.
[53] Ibid. p. 192.

For Bianchi, this role of place and material is a diminished role that is directed by the ἀρχὴ κινήσεως of the form rather than playing a co-directing role. Bianchi finds the way that form is internal for Aristotle is qua other, as a soul that is other than the body. Bianchi questions the reading of Aristotle's φύσις that I have offered that finds the ἀρχή of change to be the τέλος of change, because it is not clear how a transitive action of this source is the goal of the action. Bianchi argues that when Aristotle uses the analogy of doctoring and health he transgresses the line between τέχνη and φύσις, not so much making φύσις into a τέχνη model, but in making τέχνη into φύσις by collapsing the roles of the formal, final and efficient causes. In using the examples of medicine and the doctor as the efficient cause, and health as the form and the goal, Aristotle 'mysteriously unite[s]' the three causes other than matter 'in a grand teleological agency acting on passive matter'.[54] Later though, she will point to how health is an occasion in which matter appears capable of moving itself to its end (and interestingly, precisely through the process of heating), in a way that opens again the question of whether material moves as an aberration or if material always for Aristotle has particular powers and capacities. Bianchi argues that Aristotle 'conveniently erases' the fact that the matter in this analogy is a living body, a reading that depends on thinking that Aristotle does not see matter as powerful and living – as she writes, 'it is in the very nature of matter to be "indeterminate"' – and so concludes that to posit a living body as material is to elide something about its living rather than to elide something about material being powerful and animate.[55]

Bianchi argues that Aristotle defines matter as having no capacities of its own because it is passive in contrast to form's activity with reference to *On Generation and Corruption* where Aristotle writes that 'to be moved (τὸ πάσχειν) is characteristic of matter'.[56] But for Bianchi, the real problem is not that male is active and female passive, but that the feminine is associated with luck and chance and therefore, unpredictability. This disruptive sense of matter is what Bianchi considers its symptomatic nature. The symptom is generated by and native to the system, as the female is that which Aristotle describes as what generates in itself, not as that which orders itself but as that which is disruptive in its generativity. Bianchi makes the strong claim against those who think that Aristotle's statement that male and female differ insofar as the male is what generates in another and the female is what generates in

---

[54] Ibid. p. 197.
[55] Ibid. pp. 79, 36.
[56] Ibid. p. 48, *GC* 335b30-1.

itself softens the distinction between them that Aristotle means that male and female differ in 'body, nature, essence, and logos'.[57]

While Bianchi argues that material does have symptomatic capacities in Aristotle, she nonetheless takes Aristotle to define the female contribution, καταμενία, as prime matter, in contrast to male form. Bianchi translates Aristotle's claim that the nature of καταμανία is κατὰ γὰρ τὴν πρώτην ὕλην as 'that of prime matter'. Peck translates this phrase as 'classed as prime matter.' Platt translates it as 'an affinity to primitive matter'.[58] While material seems to be vacant of any being of its own, Bianchi reads semen's materiality as close to pure form because it is πνεῦμα, which Bianchi equates with soul heat and water, a view that suggests that 'matter as *self-moving* is indeed at work', even as she argues that in this account of semen 'its materiality [is] virtually erased'.[59] Bianchi is both pointing to ways that material has its own power in Aristotle that needs to be managed and the ways in which when it does have this power, it appears to be denied for the sake of assigning that power to more formal contributions.

Bianchi reads Aristotle's account of how the female is formed as on the one hand necessary and on the other hand the result of the failure of the male to gain mastery as a result of nature straying (this is the verb Connell considers at length, ἐξίσταται). Bianchi points to the political vocabulary Aristotle uses to describe the way that proper heat does the work to create males, while lesser (but still vital) heat will result in a female (a process that Bianchi later refers to as a 'radical transformation of something into its contrary' and claims that 'he does not argue for it').[60] How the male contribution of form should be construed in relation to female matter becomes a question for us if the apparently inert material can cause changes to the offspring through a process whereby it establishes its mastery over that principle which would appear to be defined by its mastery.

Bianchi takes Aristotle's account of the monstrosity that seems to form contrary to nature as still in a way according to nature since it somehow follows naturally that form can fail to gain mastery over the matter as evidence that forces that work against nature are associated with matter, and yet, still, Aristotle says this occurs in accordance with nature.[61] What Bianchi calls here

---

[57] Ibid. p. 35, GA 716a13-31. Deslauriers argues that the difference is nonaccidental and nonessential, 'Sex and Essence in Aristotle's *Metaphysics* and Biology', pp. 149-53.
[58] Bianchi, *The Feminine Symptom*, pp. 36, 62. GA 729a32-3.
[59] Bianchi, *The Feminine Symptom*, pp. 37, 77.
[60] Ibid. p. 61.
[61] GA 770b13-17, Bianchi, *The Feminine Symptom*, pp. 38-9.

the feminine symptom could be understood as the capacity within material for it to affect form and form's ability to reproduce itself. One way to explain this is to say that the female is an 'aleatory symptom', the placeholder for the lack of explanation at work in Aristotle's supposed account of the relation of matter and form. Another is to say that the female material, having its own capacities and ways of being, is less diametrically opposed to form, which is less anti-material than we suppose.

While Connell dismisses Irigaray's reading strategies, her own reading, though not explicitly referencing Bianchi, responds to some of the Irigarayan concerns that Bianchi's reading develops. Connell argues that the view that Aristotle is the source of patriarchal thought depends on seeing Aristotle as the source of sexist hierarchical oppositions. This position, Connell argues, depends on seeing his metaphysics as comprised of exclusive, hierarchical oppositions that constitute a rigid system which support a political and social situation that keeps men in power and has been maintained throughout history. Connell argues that oppositions can be explained in many ways, and that for Aristotle the sexes are complementary oppositions even though Aristotle sometimes describes them with hierarchical language, but more importantly that contraries do not require a rigid exclusive and hierarchical system.[62] Connell points to the Pythagorean practice of including women in their mystery rites, a remarkable fact since the Pythagoreans are viewed as the source of contraries that devalue the side of the contrary with which the female is associated. Connell denies that Aristotle associates form with reason which is opposed to irrational material in his biology. She argues that form is not associated with intellect until the medieval period, and the notion that matter is unintelligible only means that it cannot be defined, but not that it is in the realm of the unknowable.

Connell argues that Aristotle presents different accounts of matter and its relation to form depending on his specific goals of his different treatises, and does not think of the male and female contributions as 'exclusive of the properties of the other',[63] though she still resists accounts that attempt to collapse them into one another. She argues that matter is never inert for Aristotle as it is in the modern conception of matter: 'even in his crudest craft analogies, it is potent'.[64] The female material is not raw material, as earth and soil are to animals, but more as earth and soil are to plants – 'dynamic matter'.[65] Connell

---

[62] Connell, *Aristotle on Female Animals*, pp. 17–25.
[63] Ibid. pp. 4, 29, 29n33.
[64] Ibid. pp. 121–2.
[65] Ibid. p. 147.

concedes that matter is for form, even if it is always already of a certain sort. She describes Aristotle's explanations as operating top down, starting with form and function and viewing appropriate matter as what will fulfil the end, saying, 'Most matter with which a living being interacts is significant only in terms of why it is useful to its goals.'[66] She argues that Aristotle's 'dynamic and teleological model' shows the material body to be what it is for the sake of their active roles in living bodies. But she does not think that that means that matter is just 'wait[ing] around for further levels of construction to happen to it' the way it might seem on the craft analogy.[67] Against those who argue that the female contribution forms what is uniform and the male contribution forms what is non-uniform, Connell argues that the female contribution is neither uniform nor non-uniform – against pangenesis, it cannot be a formed part; but against a view of matter as mere stuff, it is also not uniform.

Her account of the material role leads Connell to conclude that, to the extent that the male contributes only form, the female has more influence on the shape and structure of the body. As Connell explains it, the materiality of the female contribution is a dynamic specific potential rather than a static structure that is at a lower level than a more complex one. 'Although the female matter does not possess form, it can be understood with reference to form; it anticipates or facilitates the advent of future form in the offspring.'[68] Noting the difference between menses and the timber in the building, Connell argues that the menstrual blood has an exclusive function, while the timber is timber as a natural thing that can also be incidentally formed into a house. As Jessica Gelber argues, the craft analogy might help understand the female contribution if it were understood as the forming of the materials for the end product, so that the work of the female is like making the brick that will become the house, but even this view would seem to make the female contribution static – completed in the bricks – though it remains dynamic and alive.[69]

While she argues that the formal principle is immaterial, Connell nonetheless maintains, 'The difference between the generative residues of the male and female is thus one of degree, undermining the idea of any mutually exclusive opposition between their roles.'[70] Both menses and semen are blood-like, Connell argues. This 'haematogenous thesis' that explains both semen and the

---

[66] Ibid. p. 136.
[67] Ibid. p. 152.
[68] Ibid. p. 156.
[69] Jessica Gelber, 'Form and Inheritance in Aristotle's Embryology', p. 202.
[70] Connell, *Aristotle on Female Animals*, p. 150.

menses in terms of blood shows how Aristotle is taking the female contribution as a starting point and needing to show that the male is like it. Since the male contribution does not look like blood, further arguments have to be made to show how it appears as it does if it originates as blood, by contrast to the Hippocratics and Galen who do not argue for a common source of the male and female contributions. This view then seems to support the notion that the male contribution is in some sense material, though Connell will reject that position elsewhere.

While Bianchi draws attention to the political language of mastery in sexual differentiation and resemblance, Connell argues that several elements of these accounts point to the power of the feminine contribution, another seed or sperm that has been mixed with the male seed in a process that determines whether the offspring is male or female and resembles the father or mother or the grandparents or ancestors further removed.[71] Drawing on medieval philosophers such as Michael of Ephesus, Connell argues that if the male can be mastered, then the female contribution must have its own δυνάμις, which suggests that the mastering of the male principle requires a female principle which masters it.[72] This δυνάμις is the capacity to loosen the movements or κινήσεις so that the resemblance goes back a generation. Connell argues that while relapse explains the resemblance going back from the father to the grandfather and so forth, only the possibility of the female potentiality overtaking the father's movement can explain the resemblance to the mother or grandmother.[73] She concludes that the passage that speaks to the failed mastery where the female can result when what is not mastered departs from type and becomes its opposite and when what is mastered departs from type and changes into its opposite refers to the same thing being mastered or failing to be mastered while different things do the failing or the mastering.[74] The female matter is what can fail to be mastered by the male qua male, having its own capacity to master as a female, so it becomes female when the male principle does not master it. The female matter can similarly be mastered by the female qua female making the offspring female.[75]

---

[71] Ibid. pp. 293–5.
[72] Ibid. p. 296.
[73] Connell follows Henry to reject Peck's translation which refers to the female and not the mother to suggest the sentence means the male can potentially produce female not that the male potentially includes the mother's characteristics, *Aristotle on Female Animals*, pp. 305–10; Devin M. Henry, 'Understanding Aristotle's Reproductive Hylomorphism', p. 274.
[74] GA 768a3–6.
[75] Connell, *Aristotle on Female Animals*, pp. 307–8.

Connell explains how the female is passive with regard to substantial change, where the female has the passive potential and the male active potential to develop a new substance, but this is not to say that one contributes more than the other to the offspring. Connell understands passivity for Aristotle as a positive capacity testified to by the account of resemblance wherein the female contribution resists, loosens and overcomes the male capacity.[76] Reading ἐνέργεια κίνησεις, actual movements, and δύναμαι κίνησεις, potential movements, in Aristotle's account of sexual differentiation and heredity as current potential and possible potential, respectively, Connell argues that the mixture that is well-concocted maintains its current potential, while the mixture that is not well-concocted due to coldness relapses or loosens, λύεσθαι, to become the possible potential, a position which is similar to Heidi Northwood's interpretation of the relevant passages that what changes over to its opposite is the fetation in a change that can be influenced by either the male or female.[77]

Connell argues that we should not suppose that there is evidence that the current potentials are superior to the possible potentials on Aristotle's theory. While Aristotle associates the current potentials with male and the possible potentials with female, Aristotle only needs to argue that one is superior, as Witt argues, not that the male is superior to the female: 'It is not the case that his general theory dictates that it would have to be the male.'[78] Connell recognises that there is sexism on Aristotle's part in assuming the well-concocted mixture produces the male who resembles the male parent, but she does not take this position to mean that any offspring that is not male-resembling-the-male-parent is a failure. Rather, Connell argues, it is deficient only in the way that the male contribution did not master the female or was mastered by the female. So Connell argues, this failure of mastery does not mean that the offspring will even lack a part, but that it is a specific deficiency relative to the individual parent and not to the species.[79] She argues further that the movements that form an individual offspring – that come from the father and the mother – shape it, but are not at work in producing form, εἶδος, a term that Connell argues Aristotle abandons when discussing heredity. Connell argues that these movements are neither form nor matter. She explains that material

---

[76] Ibid. p. 272; pace Bianchi, *The Feminine Symptom*, p. 48. Marjolein Oele, 'Passive Dispositions: On the Relationship between πάθος and ἕξις in Aristotle', Balme, 'Ἄνθρωπός ἄνθρωπον γεννᾷ: Human is Generated in Human', and Sarah Borden Sharkey, *An Aristotelian Feminism*, pp. 133–4.

[77] Heidi Northwood, 'Disobedient Matter: The Female Contribution in Aristotle's Embryology'.

[78] Connell, *Aristotle on Female Animals*, p. 317.

[79] Ibid. pp. 314–19.

is only anachronistically considered to be unshaped, even though it does not contribute form or εἶδος.

While Aristotle describes the inheritance from the mother as a departure from type, ἐξίσταται, Connell argues against translating this term as 'degenerate from one's own nature', and understands it instead in terms of change in a way that does not carry a negative or unnatural connotation. Aristotle uses the related term, παρεκβέβηκε, commonly translated as 'straying from type', to refer to the first move to female, but also to refer to the one we do not blame who is a little off the mark in *Nicomachean Ethics* and to the still fine nose that is deviating slightly from perfect straightness in the discussion of extreme versus slightly deviant democracies and oligarchies in the *Politics*.[80]

This sense of deviation is related to Aristotle's description of the female as incomplete. This view depends on seeing Aristotle's biology as having only one goal, so that every natural process would have only one goal. Connell argues that femaleness could be considered as a secondary goal for an individual animal, the lopsided cake that is still a cake, just lopsided. If animals aim for survival, flourishing and generation, the female condition is a failure only in terms of flourishing, not in terms of survival and generation. The material necessity of female births actually serves the flourishing of males, putting the female in the crosshairs of the tension between what is good for the individual and what is good for the species.[81]

Connell rejects the one-sex model in Aristotle by making the female contribution what is at stake rather than the female as such. She argues that since the mother is more fully actualised, the mother cannot be viewed as passive in contrast to the less fully actualised embryo, and thus the female contribution cannot be viewed as entirely passive. Connell argues that the female animal cannot be considered matter, because Aristotle would find it nonsensical to regard an animal as matter.[82] She argues that the female contribution cannot be viewed as a privation because the female animal would not be alive if it were a privation. So Aristotle describes the female as infirm 'as it were', with the word πῶς, which she translates as 'sort of, but not really'.[83] The female is privative in terms of the male, but on her own terms, she has her own positive function as that which generates within herself.[84] The one-sex model would

---

[80] *GA* 767b7, *EN* 1109b18–20, *Pol*. 1309b23–34, Connell, *Aristotle on Female Animals*, pp. 344–9.
[81] Connell, *Aristotle on Female Animals*, pp. 284–91.
[82] Ibid. p. 27.
[83] Ibid. p. 282.
[84] Ibid. pp. 281–4.

seem capable of accommodating this position based solely on the contributions to argue that what it is to be female is defined in terms of male who has the positive contribution to her lacking contribution. Still, the problem from the perspective of critics is precisely that the contribution is what is definitive of the female in generation to the extent that it becomes definitive of the female animal as such. The two-sex model seems to become possible when the different capacities are understood on different terms where the female's positive contribution is to generate in herself, but even then the male's contribution is an inversion of that – to contribute in another. The arguments about loosening and overcoming also seem to make generation, considered in terms of reproducing or departing from type regardless of which type is the standard, appear to reflect a one-sex model.

Connell's resistance to the one-sex model comes in the form of strongly distinguishing between the male and female principles in an effort to give proper credit to both. Connell does not think that resuscitating the role of material can mean for Aristotle eliding the fundamental difference between form and matter in generation, which is perhaps why she is ambivalent about the significance of the craft analogy in Aristotle. She argues against those views that make the work of form come through material capacities. She maintains that views that find the source of life in matter are unable to distinguish between the living and non-living. She rejects the notion that the soul comes into the offspring through the material, because she maintains that no material can pass from the father to the offspring on the basis of accepting the analogy to the carpenter. She argues that efforts to locate a material basis for form's work arise from a modern interest in making Aristotle amenable to contemporary accounts where inert material and movement are the source of what is.[85]

Yet Connell maintains that Aristotle's account of artifice cannot be equally applied to nature. She argues that Aristotle invokes the artifice analogy in order to make the case that formal and final causes are required to organise the material as a craftsperson organises the materials to make the artefact. She further acknowledges that craft and nature differ not only because nature has an internal source of change, but also because the material cause in craft is separate from the craft, though 'natural matter is present in the natural object'.[86] Moreover, natural material has its own movements and potentiality,

---

[85] Ibid. pp. 182-7. Connell's arguments against the materialist views are often arguments from motivation. For example, that the emergentists maintain their view for the sake of maintaining the primacy of chemistry.
[86] Ibid. pp. 125-6.

by contrast to materials in craft.[87] Part of Connell's argument about the craft analogy is that it should not be understood to make generation entirely due to form, since material in *both* artifice and nature is already dynamic. Connell argues that Aristotle draws on rather than rejects the Pre-Socratic tradition of viewing material as powerful. While Aristotle thinks the Pre-Socratics are too dependent on material, this critique does not lead Aristotle to strip material of its power altogether, but to foreground formal and final causes in his explanation of natural beings, which is why he invokes the craft analogy.

She further argues against taking Aristotle's craft or carpenter analogies as the most important ones when he uses other images that are equally useful like the semen as paint left on a palette when the painting is finished.[88] Connell argues that it is the female who is the automaton because matter has a specific potentiality to change or react because it is potentially the same as the female's living body. She concludes, 'Matter and form in the biological context cannot be perfectly elucidated by any comparison', dismissing the craft analogy as not giving proper place to the female. Connell argues that the analogy needs to be read in context. In some places, Aristotle's concern is to show how natural objects can have an internal source of change and yet be caused by something external, and the craft analogy is what Aristotle reaches for to explain.[89]

Connell rejects the view that everything from the elemental powers up aims to be like a god so everything that falls short is incomplete. She argues instead for a conception of teleology relative to each form. On this account of teleology, she rejects the view that makes the female a privation that falls short of completion and is merely accidental.[90] Connell's reading attempts to secure that difference of kind by making the formal contribution wholly immaterial, a reading that seems to support a craft model of natural generation. The worry is that in making form and material wholly other, the account of each of them is in securing this otherness not in finding equally positive and distinct contributions. If this otherness is what the craft analogy accomplishes, it would also seem to make the movements and contributions of material those achieved by the form and not specifically material when form is understood as what moves and makes distinct contributions.

Conclusion

The most trenchant claims made by the feminist critics of Aristotle are several. Material is passive and privative of form. The difference between male and female

---

[87] Ibid. p. 323.
[88] Ibid. 158. See *GA* 725a27.
[89] Ibid. pp. 159–60.
[90] Ibid. p. 239.

is a hierarchical difference. The ways of thinking of the capacities of menses and the work of form still make menses what it is through form without requiring a material sense of form for it to be what it is. Form, even if it is the work of elemental forces, works by mastering the material. Even the accounts that end up defending Aristotle concede a hierarchy between form and material. These are serious and textually grounded (if open to interpretation) charges that if true would make Aristotle the source of the division between material and form that maps a subservient passivity onto female and a dominant activity onto the male in a craft model where form is imposed on material. I agree with Bianchi that material has powers, but I maintain that this is a not something Aristotle is trying to repress, but part of the account that is required to make sense of how form is related to material, in such a way that offers different possibilities for seeing material in Aristotle. This account depends on focusing on the way that semen works. While Connell wants to revitalise the work of material in the tradition of Tress and others before her, she still demands a strong distinction approaching a contrary between form and matter, one that makes her sympathetic to the craft analogy in ways that raise concerns for me.

2
Disputes over the Material Contribution of Semen

The feminist critics set up a case against Aristotle that requires a turn to the specifics of generation, specifically how menses and semen work in generation. Another literature that needs to be addressed before turning to the argument of this book is the dispute over how to think about the work of semen in Aristotle's biology. This literature is divided between those who maintain that semen is a spiritual principle whose work is immaterial because of its association with soul, the so-called anti-reductionists who allow for some role for material, but argue against those they call the reductionists that the process of bringing forth life requires an explanation beyond material, and those who argue that semen's work is material and mechanical.

The crucial passage that leads to disagreement is found in *Generation of Animals* II.3, where Aristotle explains the kind of material that comprises semen to make it capable of doing its animating work:

> Now it is true that the faculty of all kinds of soul seems to have a connexion with a matter different from and more divine than the so-called elements; but as one soul differs from another in honour and dishonour, so differs also the nature of the corresponding matter. All have in their semen that which causes it to be productive; I mean what is called vital heat. This is not fire nor any such force, but it is the breath included in the semen and the foam-like, and the natural principle in the breath, being analogous to the element of the stars. Hence, whereas fire generates no animal and we do not find any living things forming in either solid or liquids under the influence of fire, the heat of the sun and that of animals does generate them. Not only is this true of the heat that works through the semen, but whatever other residue of the animal nature there may be, this also has still a vital principle in it. From

such considerations, it is clear that the heat in animals neither is fire nor derives its origin from fire.[1]

From this passage, questions are raised about what Aristotle means to say that some material is more divine, that some heat is vital or generative, that this heat is distinct from fire, that this heat is connected to breath or πνεῦμα, that the 'natural principle' is in the πνεῦμα and that this is like the element of the stars, that fire does not have the capacity found in the sun and the animals to generate life, and that not only semen, but other animal residues have generative or natural heat.

One of the central questions of this debate, and the one with which this book is centrally concerned, is: to what extent must the formal and material causes be separable in order to understand them as two separate causes? Relatedly, what sense of the formal and material causes are at work if the formal cause in generation works through material? What sense of these causes is at work if they are considered fundamentally disconnected given Aristotle's conception of natural substance as a composite of form and matter? Some who maintain that form cannot work through material go to great lengths to describe the material Aristotle describes as that through which semen works as supermaterial or immaterial. Other critics of those who call attention to the material work of semen argue that to invoke matter as the source of form's power is either to resort to pansomatism or mechanism. One view is that it is not the material composition but the form of semen that makes it form. Such thinkers insist that form must be causally efficacious as form separate from the material that constitutes the formal contribution that is semen. They argue that material might be able to do a certain amount of work, but it cannot be the cause of something with a particular function. Some accounts of materialist accounts of generation resort to something like Hippocratic pansomatism, but others argue that the source of

---

[1] GA 736b29–737a7. One quick note on the translation. I am using A. Platt's translation from Jonathan Barnes's revised edition of Ross, but I have revised it by changing some key terms in order to make some distinctions clear. I follow Robert Mayhew by translating σπέρμα as seed and γονή as semen, and vital heat as proper heat when the modifier is οἰκείου. To complicate things, Lesley Dean-Jones translates γονή as both seed and sperm and σπέρμα as semen in the Hippocratic corpus; Dean-Jones, *Women's Bodies in Classical Greek Science*, pp. 154, 155n25, 155n26, 165n60, 166n61. James Lennox argues that Aristotle uses σπέρμα somewhat interchangeably with γονή in *GA* I; in *GA* II and for the rest of the book, Aristotle seems to use σπέρμα generic for seed or contribution and γονή to refer to the male contribution or semen and καταμενία to refer to the female contribution or menses. James Lennox, 'Aristotle's Biology'.

the distinction between form and matter is due to differences in motion between them, raising the question of the extent to which motion can produce a difference between form and matter. These questions and these disagreements lead back to arguments between Plato and Aristotle and their reception about form and the extent to which species form exists independently, which I consider in the second part of this chapter.

Robust Material View versus Restricted Material View

The dispute between those who aim to explain Aristotle's account of generation through its material processes and those who argue that Aristotle's metaphysics demands an explanation with reference to formal and final causes has been framed as a dispute between what Sheldon Cohen calls the reductionist and the anti-reductionist views. This language strikes me as already favouring the anti-reductionist position, so I am renaming it the robust material view against the restricted material view. The restricted material view is that heat and cold can produce flesh and bone, but they are insufficient to cause living substances. Heat and cold can explain individual changes, but not the series of changes that lead to substantial change. Cohen takes this position further to argue that heat and cold cannot even sufficiently explain individual changes, but only passive potentialities.[2] The robust materialists' challenge, according to the restricted material view, is to explain not just that these events take place, but that this series of events takes place without referring to form.[3] Cohen does not suggest that the restricted materialists have a challenge, but they do, which is to explain how form can materially work on matter if it is wholly other than matter.

One problem with this way of articulating the dispute is that it determines in advance where the work is material and where it is formal, what the work of matter is and what the work of form is, which I maintain is called into question by Aristotle's account of generation. Another problem is that Cohen focuses his account on passages that address the work of heat and cold in forming parts without examining the passages where Aristotle explicitly connects heat and cold to the work of form in generation. The dispute is to what extent the material and formal explanations can be separated. To what extent is generation due to the necessary characteristics of the material at work and to what extent is generation due to the formal cause and the τέλος to which it aims? To what extent are these causes divisible in natural generation and natural substance?

Friedrich Solmsen argues that Aristotle treats vital heat as 'a new discovery' of a 'peculiar substance' (a translation of σῶμα, commonly translated as

---

[2] Sheldon Cohen, 'Aristotle on Heat, Cold, and Teleological Explanation'.
[3] Allan Gotthelf, 'Aristotle's Conception of Final Causality', pp. 208–12.

body, from the passage above in which Aristotle distinguishes vital heat from fire).⁴ On the basis of the passage quoted above, Solmsen argues that this body is not fire, but vital heat and inborn πνεῦμα; it is analogous to αἰθήρ. Solmsen argues that πνεῦμα is 'a more "spiritual" principle' than a material one, and necessarily so in order for it to be capable of being the source of soul. He maintains that αἰθήρ is divine and unchanging material, and bases his reading on Aristotle's non-extant dialogue *On Philosophy* and the first book of *De Caelo*, where Aristotle discusses an ungenerated and indestructible element that comprises the heavenly bodies.⁵ Solmsen finds in this element a material-like stuff that can do the work of form without having to make form dependent on material. Solmsen argues that *Generation of Animals* I 'has assured us that the male parent contributes nothing material to the foetus but only εἶδος and ἀρχὴ κινήσεως'.⁶ Solmsen concedes that 'the *sperma* must contribute something material, albeit the finest and noblest material, a φύσις analogous to the aether'.⁷ Aether then functions as a material that is more formal or spiritual. For this case to work, Solmsen argues that Aristotle raises the πνεῦμα in which the αἰθήρ is found above the level of wind or breath, as it was commonly used, to something more divine.

Solmsen argues that Aristotle's vacillation between speaking of these as material and at other places as immaterial or divine points to his need to find a material basis for soul while resisting a reduction of soul to material. Tress similarly argues that vital heat can explain what ordinary heat cannot – how matter can be organised – which is to say, it can do the work of the form and final cause. Tress argues that 'a special principle' is at work in the πνεῦμα that explains its constitutive role in semen to bring about life in the offspring. She takes specific aim at those who think the special principle in πνεῦμα is merely hot air.⁸ On Tress's account, the material explanation is insufficient, even in places where it can explain how heat develops, because the heat that is the

---

⁴ Friedrich Solmsen, 'The Vital Heat, the Inborn Pneuma and the Aether', pp. 119-23.
⁵ *De Caelo* 270a13-270b30. Aristotle's support for the immortality of αἰθήρ turns to a discussion of the immortality of the gods who occupy the highest heavens, which is an odd turn for Aristotle. In any case, the strongest case seems to be that its movement is circular and it is without a contrary, thus it cannot move in a way that would lead to destruction or generation.
⁶ Solmsen, 'The Vital Heat, the Inborn Pneuma and the Aether', p. 121.
⁷ Ibid. p. 121. See *GA* 727a35-b1.
⁸ Daryl McGowan Tress, 'Aristotle against the Hippocratics on Sexual Generation: A Reply to Coles'. Tress is here arguing against Andrew Coles's interpretation of generation in Aristotle that attributes a pansomatist view to Aristotle in order to give material a central role. Tress argues by introducing the formal and final causes that Aristotle distinguishes his account from the pansomatist accounts of the Hippocratics. Abraham P. Bos similarly argues that πνεῦμα is not merely air in *Aristotle on God's Life-Generating Power and on* Pneuma *as Its Vehicle*, p. 143.

result of material or mechanical movement would still need a formal or teleological intervention to do its work. Lesley Dean-Jones similarly suggests that Aristotle maintains a distinction between the form and the matter in order to keep the male and female contributions distinct. Dean-Jones argues that Aristotle's unwillingness to allow for a material role of semen in the newly formed embryo 'shows that the separation of form and matter was a very strong principle in his theory of reproduction'.[9]

Yet Solmsen locates a tension in Aristotle between the need for material and the divine source of life, which Solmsen contextualises in the thought of Aristotle's predecessors and contemporaries, concluding that Aristotle's αἰθήρ is like the principle of air the Pre-Socratics associate with soul.[10] What he finds significant here is that from Empedocles through Plato and then Aristotle, air is one of the elements that comprise living substances. Aristotle's innovation is to break with this tradition when he associates vital heat with πνεῦμα and almost but not quite with αἰθήρ, a fifth element.[11]

This effort to associate the materiality of αἰθήρ and the celestial bodies can be traced to neo-Aristotelian and Arabic philosophers. Theological vitalism views the male role as divine, the material source of form in semen, αἰθήρ in the πνεῦμα, as itself divine. This view arises from Arabic thinkers like Al-Farabi, Avicenna and Ibn Bâjja, who argue that form as something above matter means for Aristotle that form emanates from divine intellect.[12] The view that the male

---

[9] Dean-Jones, *Women's Bodies in Classical Greek Science*, pp. 187–8.

[10] Solmsen notes that 'all things are full of soul' replaces Thales's 'all things are full of gods'. He traces a line from Anaximenes, the Milesian pre-Socratic philosopher who argues that everything is air, to Diogenes of Apollonia, who defines σπέρμα as foam. Cf. Rhodes Pinto, '"All Things Are Full of Gods": Souls and Gods in Thales'. Anaximenes argues that 'just as our soul (ψυχή), being air (ἀήρ), holds us together, so do breath (πνεῦμα) and air encompass the whole world', DK13B2, *A Presocratics Reader*, p. 20. Diogenes of Apollonia takes this further by arguing that air itself, which comprises all things, is intelligent. Air reproduces itself through aerated seeds. Living things are then generated through aerated seeds (σπέρμα), which achieve a level of intelligence when the blood becomes aerated, or, when the air is internalised in the blood in Diogenes of Apollonia Fr. 5 and DK64A19. See Kirk, Raven and Schofield, *The Presocratic Philosophers*, pp. 442–9. Solmsen notes that Aristotle's πνεῦμα is inborn by contrast to Diogenes' concept. He references scholars who trace Aristotle's view to Diogenes, while others trace it to the tradition of Empedocles. On the side of Diogenes, M. Pohlenz, *Hippokrates und die Begrundung der wissenschaftlichen Medizin*, pp. 39ff., 93ff.; Erna Lesky, 'Die Zeugungs- und Vererbungslehren der Antike und ihr Nachwirken', pp. 19, 123ff. On the side of Empedocles, W. Jaeger, 'The Pneuma in the Lyceum'.

[11] Solmsen, 'The Vital Heat, the Inborn Pneuma and the Aether', pp. 119–23.

[12] This view depends on an error in translation which deletes the μέν ... δέ/on the one hand ... on the other hand structure from *GA* 737a8–16, which leads the passage to read that seed is always separated from matter. See Sophia M. Connell, *Aristotle on Female Animals*, p. 189.

role is superior and the female role is 'earthly and degraded' follows from medievalists like Albertus Magnus, who is the source of the view that if male is responsible for the soul in the embryo, then the intellect is transmitted by the male through the semen.[13] Albert nonetheless maintains that forms exist in material, rather than as immutable ideas.

Against the view that the intellect or a more spiritual substance is the source of life, some scholars focus on the movement of semen as the cause of generation. D. M. Balme argues that Aristotle's explanation for the need for sexual differentiation in reproduction that is based on something better and for the sake of something is based on a *prior* principle, rather than a principle from above, or the prime mover, as Solmsen has argued.[14] Balme reads ἄνωθεν, which is literally 'above', as prior, arguing that Aristotle regularly uses this term to refer to prior members of the species, so it seems to appeal to a more general final cause, rather than something that transcends material substances altogether.

If the separation of the male from the female is not because the male principle is more divine, Balme argues, a 'mechanical cause' can be located in the movement in the semen, which continues after the semen leaves the father's body, which means that the movement belongs to the semen and eventually belongs to the offspring. The semen works 'to initiate such movements in the material as will transform it into a growing embryo' by causing the heart to be formed through the semen's movements.[15] Πνεῦμα is the cause of this movement because it conveys the heat that has the capacity to organise the female material into a perceiving animal. When semen imparts its heat to the foetus – Balme says it loses its heat to the foetus – at this point, the material of the semen evaporates, being deprived of the heat that kept it moist. While the moisture has evaporated, the heat from the πνεῦμα remains and is able to continue to convey soul, both the perceiving and rational soul. Balme maintains that the different gradations of generative heat in the πνεῦμα produce different degrees of complexity of soul from the nutritive soul to the rational soul.

Balme argues that Aristotle does not mean 'something novel' by his claim that πνεῦμα has a generative heat.[16] In Aristotle's time, Balme notes, πνεῦμα meant wind. When associated with animals, it meant breath or wind, and it seems to originate outside the animal. This πνεῦμα is generally understood akin to breath and air, yet while air is composed of the hot and the wet, outside of

---

[13] Albertus Magnus, *On Animals*, II.16, 12, §65.
[14] D. M. Balme, *Aristotle's* De Partibus Animalium *I and* De Generation Animalium *I*.
[15] Ibid. pp. 155-8.
[16] Ibid. p. 161.

Generation of Animals its function is to cool rather than to heat. Only here does πνεῦμα take on the capacity to heat.

Notably, this reading makes πνεῦμα something mysterious in generation, not unlike Solmsen's account, yet Balme distinguishes his reading from Solmsen's in important ways. Balme argues that it is not πνεῦμα but generative heat which Aristotle associates with αἰθήρ. Yet after making this association, Aristotle does not return to the πνεῦμα again. Balme distinguishes between the nature that is in the πνεῦμα from the nature of the πνεῦμα to point to how the significant element of the πνεῦμα is the heat.[17] It is not entirely clear whether the πνεῦμα or the heat is what varies in degree to produce different souls. Moreover, Aristotle does not identify the generative heat directly with αἰθήρ. While he calls the material at work in semen divine, he is making a comparative claim that it is more divine than fire, which means, according to Balme, that it is 'less grossly material, purer, superior'.[18] Balme concedes to Solmsen the need to make semen work through something less material, but he nonetheless concludes from πνεῦμα or breath being formed when heat acts on moisture as a result of the body's natural activity that insofar as πνεῦμα is the source of heat in semen, there is a continuum here between heat and vital heat.

Distinguishing heat from vital heat, Balme similarly distinguishes πνεῦμα in breathing from the πνεῦμα in the seed that contains generative heat. Balme explains that heat that generates life shares a capacity with the sun that it does not share with fire. Balme argues that 'a state of superior "purity"' enables it to convey soul to the offspring. Especially rational soul is attributable, according to Balme, to 'the purity of heat in his heart'.[19] Balme's account is material insofar as he argues for a continuity between the non-living and the living, and thus for a continuity between other heat and generative heat, between atmospheric air and πνεῦμα, which is converted by animal heat to enable it to animate further life when carried by the semen. Like Solmsen, Balme connects divine material with purity – here drawing on the association of αἰθήρ with the divine in *De Caelo* I – to explain the *scala naturae* and the generative capacity of heat, but he presents a challenge to those who would say that Aristotle's account of generation through heat contradicts his accounts of soul or of matter elsewhere.

Like Balme and Solmsen before him, Freudenthal points to a special kind of material, which he calls connate πνεῦμα, in the sublunary level, that reflects

---

[17] Connell argues that this distinction is what shows that Aristotle is not analogising between αἰθήρ and πνεῦμα in his discussion of vital heat but between what makes each hot in *Aristotle on Female Animals*, pp. 190-1.

[18] Balme, *Aristotle's* De Partibus Animalium *I and* De Generation Animalium *I*, p. 163.

[19] Ibid. p. 164.

the more celestial element of αἰθήρ. But Freudenthal's account explains how the connate πνεῦμα comes to have an enforming capacity through a mechanical process of pneumatisation in the blood. Pneumatisation produces pulsation or movement in the blood, which makes the work of pneumatisation like vapour. This πνεῦμα remains in the body just as blood does. Freudenthal argues that πνεῦμα remains in the blood in a similar way to how milk becomes airy through the process of heating. As Aristotle explicitly says, blood is not essentially hot, but becomes hot and carries vital heat throughout the body from the heart, which is the internal source of heat. Freudenthal argues that vital heat is transported in the blood through connate πνεῦμα. If things become better in the body by moving up, or move to the heart through moving up, the problem is that vital heat cannot move up on its own, since it is not an element. But connate πνεῦμα, being hot as fire is, can move up. He argues further that sensations are also carried in πνεῦμα, which does the work of maintaining material persistence of substances. Thus, connate πνεῦμα does on the sublunar level what αἰθήρ does on the celestial level.[20]

Freudenthal maintains that this account of a material cause that is the source of both material persistence and of metaphysical distinctions of soul is true of the biological works, but inconsistent with later metaphysics and closer to the Pre-Socratics.[21] Freudenthal interprets Aristotle's claim that vital heat is the heat from the sun and from animals in a materialist vein with reference to the passage in *Generation of Animals* III.3 where Aristotle explains spontaneous generation in order to show how matter with no prior history of being alive can receive a heat from the sun that makes it live.[22]

Following Freudenthal's effort to present a materialist view of vital heat, Paul Studtmann argues that vital heat is the biological mechanism that explains the differences between souls. In contrast to Freudenthal, Studtmann argues that vital heat is the heat that causes matter to emanate heat rather than to move simply upwards or downwards. As heat internalised in matter, the heat and the matter would have a natural limit because at a certain point, it would cease to emanate because it had become cold. The heart, Studtmann argues, is the source of this internalised emanating heat and therefore also of growth.[23]

Animals need nourishment. Nourishment is accomplished by concocting food through heat. Food is converted by vital heat in the stomach. The food

---

[20] Freudenthal, *Aristotle's Theory of Material Substance*, pp. 136–41. Bos makes a similar argument in *Aristotle on God's Life Generating Power*, p. 206.
[21] Freudenthal, *Aristotle's Theory of Material Substance*, p. 87.
[22] Ibid. p. 26. *GA* 762a18–31, b9–18.
[23] Paul Studtmann, 'Living Capacities and Vital Heat in Aristotle'. Studtmann references *Juv.* 467b14–16, 468b28–469a6, p. 369.

turns into blood and then semen in the male, which can produce perceiving organisms, and menses in the female, which can begin the generating process. The vital heat causes motion in the semen, which carries the male form in actuality and the female form in potentiality, so that if the motion is incomplete, the potential overcomes the actual.[24] The heart becomes the emanating source of heat in the offspring, at which point it has its own capacity for growth. Thus, the same degree of vital heat explains the capacity for nutrition and reproduction.

Studtmann proceeds to explain that perception and desire depend on organs that have certain ratios, which he speculates are maintained by a greater degree of vital heat than is required for nutrition and reproduction. Studtmann argues that objects of sense-perception are ratios because physical objects are the result of mixing of elemental forces, suggesting that form is equivalent to the material mixture. From this point, Studtmann claims 'it is a small step to the idea' that sensible objects can be divided between those that are commensurate with the ratio of sense organs and those that are not, and a further small step to the claim that the commensurate objects are pleasant and the incommensurate ones are not pleasant.[25] What it means for a sense object to be incommensurate to a sense organ would seem to mean that it would not be perceptible to that organ, on Aristotle's terms, the sense object is the actuality of what the sense organ is in potentiality, so a sense object that is not commensurable with the sense organ, would not be the actuality of what it is potentially and so would not be perceived.[26]

Freudenthal and Coles make much of the movement in Aristotle's biology as the source of the vital heat which becomes the source of life in the offspring.[27] This account follows the view that generation occurs analogously to puppetry, as Connell argues. She describes two types of ancient puppets, the wind-up toy where one part moves another which moves another and so forth

---

[24] This view that the semen carries male actually and female potentially would seem to subject Studtmann to the preformationist position, which Aristotle explicitly rejects, where the semen has parts actually within it, an account which leads to the further difficulty of how the actual is overcome by the potential in Studtmann, 'Living Capacities and Vital Heat in Aristotle', pp. 371–2.

[25] Ibid. p. 377.

[26] *De An.* 424a18–29. Aristotle does conclude in this passage that 'excesses in objects of sense destroy the organs of sense', but here the point is that the object can be excessive, not that it is incommensurate with the organ.

[27] Andrew Coles, 'Biomedical Models of Reproduction in the Fifth Century BC and Aristotle's *Generation of Animals*', pp. 45–88.

and the kind where one movement creates instantaneous action in the parts. The second would seem to be closer to what Aristotle means because with it, some work of the soul or form causes the shaping of the offspring in such a way that produces the capacity for the offspring to organise itself.[28]

Connell argues against functionalist accounts which take the material cause to be sufficient for explaining the work of form because, while they allow for the macroscopic account that Aristotle offers with four causes, they make the formal and final causes 'causally inert'.[29] She maintains that form must be causally efficacious and immaterial.[30] Connell criticises the material accounts of semen, maintaining that it is not through what semen is made up of but through the form of semen that soul comes into the offspring.

Connell calls the argument that makes matter responsible for life the 'hylozoic argument'. She argues that on this account there can be no difference between the living and the non-living, and that living things only differ from non-living by containing special living materials. She rejects this view because it would require special material passing from the father to the offspring. Connell rejects this possibility with reference to Aristotle's claim that the material of the semen evaporates and that since Aristotle maintains that no material passes from the father to the offspring, material plays no role in animating menses.[31]

For Connell, the accounts that make the male role mechanical – based on movement caused by heat – 'mar our understanding of the female role', which she argues can be said to do more and more independent work if the male is not doing this work through heat. Against those who say that the male role is divine and those who say it is material or locomotion, Connell argues that it is immanent immaterial form. The male contribution is a principle of movement that joins the efficient and formal cause, in a way that makes generation look more like building than, as I will argue in the next chapter, teaching in which an agent works on a patient that is unlike in order to make it like. While Connell argues for a methodological approach that takes *Generation of Animals* on its own terms, she argues that generation cannot occur through heat

---

[28] Connell, *Aristotle on the Female Animal*, pp. 224–7.
[29] For functionalist accounts, see Martha Nussbaum, 'Aristotle on Teleological Explanation', and Monte Ransom Johnson, *Aristotle on Teleology*, p. 182.
[30] Connell, *Aristotle on Female Animals*, pp. 183, 197.
[31] See GA 730b10–15, 737a8–15. Note that in other contexts, Connell is willing to stress the disanalogy between nature and craft, as when she argues that nature has an internal source of organisation in contrast to craft and when she warns against taking it too seriously, *Aristotle on Female Animals*, pp. 121, 159–60, 176n42.

because otherwise the change would not be substantial. This methodological inconsistency that privileges an account of substance based on arguments from elsewhere in the corpus emerges when Connell rejects the notion that πνεῦμα can be the cause of the formal nature of the semen on the basis that the mother also has some degree of πνεῦμα, arguing that the difference between the male and female contribution must be a formal difference. Connell argues that the view that material explains the work of soul in semen must be rejected because this view is 'at odds with Aristotle's outlook in biology', a strange claim given that this passage would seem to be central to determining that outlook, not something that can be determined *a priori* from the earlier chapters.[32]

Connell argues further that the vital heat explanation for how semen works is not necessary and sufficient insofar as it explains both animate and inanimate processes, such as when milk is curdled in cheese-making. While Aristotle calls this heat natural heat, soul heat, life heat, Connell argues that the comparison of generation to cooking in which heat does the work is to point to the work of the chef rather than the vitality of the heat, since the chef is what makes the heat have its effect, an interpretation that does little to distinguish between vital heat and any other kind of heat. Connell rightly points to the question of how vital heat imposes a limitation, which is the work of the animal soul.

Connell occupies an interesting position between the spiritual account of the work of semen and the material or mechanistic account, rejecting theological vitalism for similar reasons that she rejects hylozoism: the work of one particular material is taken to be a special kind of work. Theological vitalism depends on a reading of αἰθήρ in Aristotle that takes the work of πνεῦμα to be analogous to the work of αἰθήρ. Connell says that Aristotle does not maintain that πνεῦμα is analogous to αἰθήρ, but that what makes the semen fertile or hot is analogous to αἰθήρ.[33] While Connell denies the divine connection to αἰθήρ that the neo-Aristotelians and medievalists find, she also denies the view that πνεῦμα can do the material work of nature because she argues there is no systematic account of πνεῦμα to be found in Aristotle. Connell cautions against seeing Aristotle as the

---

[32] Connell, *Aristotle on Female Animals*, p. 263. On the same subject of how the πνεῦμα works, Connell writes: 'Although he says the soul principle emerges in the enclosed *pneuma*, this counts as a somewhat feeble attempt to describe what is taking place rather than any positive endorsement of the idea that all generation is down to a special material or a mechanical process.' Such a view depends on privileging the earlier more general chapters and then interpreting the later chapters in an effort to find consistency with the more general chapters, a method Connell expressly rejects when she writes, 'Thus, I will urge that the *GA* ought to be read as a whole, with equal emphasis placed on each part rather than using the technique of subordination, and in doing so problems with coherence and consistency are greatly diminished', p. 56.

[33] Ibid. pp. 188–94.

conceptual link between the Pre-Socratics and Hellenic philosophers for whom heat and breath is central, but another way to understand it is not that Aristotle is a conceptual link but that they all share a context in which the work of heat is taken as background and the specifics of how it works is what is in question. Both Stoics and Hippocratics, as Connell notes, see πνεῦμα as the material substance of the soul, but Connell's account of vitalism 'posits a fundamental ontological distinction between matter and life. For Aristotle, for example, matter is not able to be alive without the presence of soul, which is the source of life and is irreducible to matter.'[34] Connell argues that the account that explains soul through the air warmed by vital heat to form πνεῦμα requires that we systematise what is not systematic in Aristotle's account.

Connell argues for a vitalism of a form that works in the natural world separate from material. She rejects the view that heat as a material power can make a difference for causing life and in sex differentiation. She rejects the view that the parts – the testes or the uterus – that conduce to more or less heat make the difference.[35] As in her discussion of soul and its relation to vital heat, Connell separates the principles from the parts that do their work instead of finding them interwoven within them. She concludes that the principles of maleness or femaleness precede the organisation of the body and determine how it is structured so that the movement is from function to parts to temperature and not vice versa.

Connell and Tress reject the view that semen does some magical work, because neither of them think that ultimately, the material of semen is doing the work. Both of them are left with the question of how form works. Connell's interpretation of male as the ἀρχὴ κινήσεως is that the male is the principle of a transition, something over and above the heat that forms the parts but cannot form them as parts of living bodies. The movement of the male is the movement that begins a substantial process of change.[36]

Navigating between the formal and the materialist positions, Dean-Jones argues that a tool, semen, is required to transfer the formal principle to the material, a tool that she names the efficient cause.[37] But, she explains, semen itself is dependent on the material of the residue in order to become semen. This dependence explains why those who are fat produce less semen than the lean, 'The reason is that fat also, like seed (σπέρμα), is a residue and is in fact

---

[34] Ibid. p. 218.
[35] Ibid. pp. 268–70. Contra Marguerite Deslauriers, who says the parts make the difference because they are the source of heat in Deslauriers, 'Sex and Essence in Aristotle's *Metaphysics* and Biology', pp. 148–52.
[36] Connell, *Aristotle on Female Animals*, pp. 161–70.
[37] Dean-Jones, *Women's Bodies in Classical Greek Science*, p. 184.

concocted blood, only not concocted in the same way as the seed (σπέρματι). Hence it is not surprising that when the residue has been consumed to make fat the semen (γονήν) is deficient.'[38] If the residue is consumed to form fat the semen is naturally deficient because there is not enough residue to form semen, making semen dependent upon the material available for it to come into being.

Balme explains the problem of *Parts of Animals* in a way that captures the robust material versus restricted materialist debates in Aristotle's biology: '[H]ow is teleology to be reconciled with material-efficient causation, is the final cause really an objective factor or is it only a heuristic device of the biologist?'[39] Continuing his argument in a later essay, Balme argues that the form is conveyed by the male parent, who forms the embryo through the movement of the semen, while contributing 'nothing somatic', by which Balme seems to mean, nothing material.[40] Balme argues that this explanation shows how form and finality work at the physical level, which is an answer, Balme argues, to the problem Aristotle raises in *Metaphysics* VII of how the individual natural living substance can be definable insofar as it is material.[41] Balme argues that his account shows three things that point to how form is individualised in the work of generation: how a movement can continue after the mover ceases to work on what is moved, how a movement can be actually simple while potentially complex, and how 'a complex of movements can control and direct itself'.[42] Balme argues that there is no further material influence yet this work does individualise, concluding that not matter but the motions of form do the work of individualising, and that the parent aims to preserve its own form, which it does by passing it on.[43] Thus the form and the final cause are explained by the act of generation that is not external to the act.

Material Necessity and Teleology

One argument made by those who reject the causal and explanatory power of material in natural generation is that *Meteorology* IV.12 establishes that material cannot explain the purpose and hence is not itself for a function. Gill enters this debate with a reading of *Meteorology*, the 'bridge between [Aristotle's] works on

---

[38] *GA* 727a34–727b1.
[39] Balme, 'The Place of Biology in Aristotle's Philosophy'.
[40] Ibid. p. 10.
[41] Ibid. p. 11.
[42] Ibid. p. 18.
[43] Ibid. p. 19. John M. Cooper similarly defends the view that form in generation is individual in 'Metaphysics in Aristotle's Embryology'. More recently, Gregory Salmieri defends the view that form in generation is a robust particular. Salmieri argues that the mother's material movements as well as the father's formal movements contribute to form in 'Something(s) in the Way(s) He Moves: Reconsidering the Embryological Argument for Particular Forms in Aristotle', p. 194.

inorganic matter and his works on biology'.[44] The book as a whole focuses on matter one level up from the elements, which are treated in *De Caelo* and *On Generation and Corruption*. The focus of *Meteorology* IV is the stuff composed of the elemental – the uniform homoeomerous bodies. The first eleven chapters explain the generation and composition of these parts without reference to the substances that they constitute. The discussion is free of teleology and focuses instead on 'the actions and effects of materials behaving according to their natures'.[45] The last chapter of *Meteorology* IV explains how we can fully understand the uniform stuffs when we examine both form and matter as the causes of generation and change. Gill argues that this shift to both matter and form, after eleven chapters of describing the work of material, should not be understood as dismissal of the power of matter and an argument for universal teleology that demands everything be explained in terms of formal and final causes, but a 'propaeudeutic' to the biological works. The teleological claims that reach down to the elements in this chapter are limited to material components of natural substances and artificial objects. Material explanations can still be given for uniform parts and elements beyond those things. Gill argues that in these cases, the material is best explained in terms of the material forces that make it what it is. If the elements are the material end of the scale, and substance the formal end, the elements that constitute an organism are what they are with reference to the form. The form proper to a functional part is its functional form, and material parts have functions that are less clear with reference to the substance the farther they are from substance. Fire cannot be the cause of nutrition, because fire is not its own limit, the soul is, so 'the functional-form of fire within an organism has transformed the element fire into vital heat: the heat concocting food into blood, and blood into flesh and bone, is no ordinary fire but fiery psychic heat'.[46] On this reading, fire becomes vital heat through its service to form. Certain substances need to be made of certain materials in order to fulfil their function. Fire becomes the heat needed to bring the soul into the female menses in the service of the form of life. Form transforms fire by making it in the service of life, which is where form seems to have a metaphysical force. Yet what has preceded this discussion in *Meteorology* IV explains how the material itself can do the work necessary to make fire into the vital heat that the form requires. Gill acknowledges this possibility when she explains the production of 'causal and dispositional properties' of elemental materials by material necessity, that is, according to their material natures. As Gill writes, 'These properties emerge through the action of heat and cold on elemental materials', which is the site of an 'ineliminable "bottom up" story', which is 'not

---

[44] Mary Louise Gill, 'The Limits of Teleology in Aristotle's *Meteorology* IV.12'.
[45] Ibid. p. 337.
[46] Ibid. See *De An.* 416a10–18.

only about inanimate material objects but about living organisms as well'.[47] The bottom up story gives power to material on its own terms. The bottom up story is that the body needs material that necessarily acts in a certain way on other material in order for it to animate it, because form is not an external organising force, but internal, and as internal works through material. The way that semen works as form through material, and thus, between necessity and teleology, will become a concern again in Chapter Five.

Aristotle and Plato on Form

The question of the extent to which the material composition of semen contributes to its function as form heralds more metaphysical arguments about how form works in Aristotle and the extent to which form and essence are distinct from the composite substances they form. The question of whether semen can work in and through material is the question of to what extent natural form is not only enmattered in natural substance, but enmattered qua form, a position that requires form to be individual. Aristotle explains the material process through which semen works. The role of material in semen prompts the question of the extent to which natural form as a principle is enmattered. The argument that allows form to be enmattered, but as the agent of species form, highlights the continued influence of Neoplatonism in interpretations of Aristotle. These readings establish a distance between form and the individuals in which form is instantiated in a way that makes form have more ontological reality than individual substance.[48] The case against species

---

[47] Gill, 'The Limits of Teleology', p. 348.
[48] The debate over whether form is universal or individual remains unresolved in Aristotelian scholarship. Mary Louise Gill argues that the generally accepted interpretation argues from *Metaphysics* VII.13 that form can be a universal and be substance, it just cannot be universally predicated, a view that can be traced to Michael J. Woods, 'Problems in Metaphysics Z, Chapter 13'. Michael J. Loux argues that form is universal, and predicated universally, but not predicated universally of the individuals whose substance it is in 'Form, Species, and Predication in Metaphysics Z, H, and Θ'. See Gill's excellent overview of these problems in 'Aristotle's *Metaphysics* Reconsidered', p. 233. The extent to which one views this as the generally accepted interpretation depends on whether you accept that interpretation. James H. Lesher rejects the view that Aristotle distinguishes between a universal and being predicated universally because Aristotle uses καθόλου and τῶν καθόλου λεγομένων interchangeably in Lesher, 'Aristotle on Form, Substance, and Universals: A Dilemma', pp. 170-1. The argument that form is individual in Aristotle can be traced to Rogers Albritton, 'Forms of Particular Substances in Aristotle's Metaphysics', and is developed by Michael Frede and Günther Patzig, 'Sind Formen Allgemein oder Individuell?'; Terence Irwin, *Aristotle's First Principles*, Ch. 12; Charlotte Witt, *Substance and Essence in Aristotle: An Interpretation of Metaphysics*; Jennifer E. Whiting, 'Form and Individuation in Aristotle'.

form and thus against the argument that material is strictly a cause of individuation or that form is such a cause in substance without reference to material will show that form works directly in and through material and not, as the advocates for species form and the Neoplatonists suggest, at a considerable distance from it. Semen is the power of form, not at a remove, but in the work that it does as the enmattered organising principle of the individual. Material is not only part of the definition of what it means to be an animal of a certain kind, but also that whereby the form of the animal works.

The question of whether form is individual or species form arises in Aristotle's accounts both of how form works through the material capacities of sperm and how form works in the process of inheritance. In the latter, Aristotle suggests that the failure of the father's contribution to concoct leads to a loosening of the capacity of the male to produce inheritance, a loosening which can recede to the point where only the universal resemblance is achieved. Some scholars have concluded from this claim that the species form not the individual form is causal, a position which supports the view that the work being done in generation is not by the enmattered semen, but by a transcendent form.[49] Chapter Six addresses this passage on failed inheritance of traits to show that the failure of individual resemblance does not necessitate species form. Against the argument that there must be separate species form in order for there to be resemblance, the reason that resemblance occurs is that this form comes from and becomes individuated from a substance that is of the same form. Form preexists the generated natural substance because a substance of similar form exists as the parent.[50] If universal form exists it is a certain sense of the definition or formula, not distinct from form, since the definition is of the form, but form does not exist separately from the existing individuals. This section defends the view that form is individual in order to support the argument that semen works through the material that constitutes it, being the power of form and not just a carrier of the capacity for form. While this reading draws from the disputes in the literature, it comes to the further conclusion that the arguments for enmattered individual form require a reconsideration of the meaning of matter and form that allow them to be interwoven.

The process of generation and inheritance occurs through the work of enmattered form – semen – form individuated from the father's form and the power to cause the offspring. Like Plato, Aristotle argues that substance is individual, and form is causal of what is, but not in a way that is one-over-many,

---

[49] Gelber, 'Form and Inheritance in Aristotle's Embryology'.
[50] *Metaphysics* 1034b10–15, but see 1034a2–5, 21–30, a35–b8.

but one in one. Form for Aristotle does not explain identity or the necessary properties of a substance but the being of substance, as Charlotte Witt has argued.[51] For this reason, form must be individual, because what is in common cannot be causal of individuals. When essence is understood to be causal of the individual substance, and definition is of the cause, then the individual substance is ontologically and epistemologically prior.

Aristotle presents a dilemma for both those who would deem form individual and those who conclude it is universal, because what most is – substance – must be knowable and a this, yet only universals are knowable. The traditional approach to reconciling these positions attempts to show that universals can be substance or form is a substance in a very specific way for Aristotle.[52] But another way to approach the dilemma is to reconsider the view that the individual is unknowable. Instead of seeing definition as common, if definition is a statement that indicates essence, and applies most to individual separable substances, then individual essences can be defined.[53] The view that knowledge must be universal and thus its object – form – must be universal to be knowable is taken as a truism, but Aristotle distinguishes between actual knowledge that is definite and of a determinate being and potential knowledge that is universal and indefinite in *Metaphysics* XIII.[54] The potential knowledge would be of what is in common between individual forms, but this knowledge itself would not be essence, which is causal and so not common.

On the traditional view that considers form species form, material operates as the principle of individuation, not knowable in itself, but causal of the differences between members of the same species. For material to be the cause of individuation of ontologically basic substances, material would have

---

[51] Charlotte Witt, 'Aristotle's Essentialism Revisited'. Alan Code argues that substance is what makes a thing what it is and is not a property in 'The Aporematic Approach to Primary Being in Metaphysics Z'.

[52] See Lesher, 'Aristotle on Form, Substance and Universals: A Dilemma'. Variations to reconciling this account of the dilemma are discussed in the first section of this chapter, and include in addition invoking a sense of individual as predicable of itself that makes form both individual and knowable because held in common and thus universal; see Code, 'Aristotle: Essence and Accident'. Support for these positions is developed from *Meta*. 1038b8–12, 1034a5–8, *GA* 730b35, *Meta*. 1036a28, 1032b1–2, 1033b17, 1037a27–1037b7, 1040a5–7, 1041b4–8, 1050b2.

[53] *Meta*. 1030a17–1030b5.

[54] See Frede and Patzig, 'Sind Formen Allgemein oder Individuell?' p. 56 and Witt, *Substance and Essence in Aristotle* VII–IX, pp. 168–78.

to relate species form and material in a fundamentally unified way, which is to say, material would have to act as individual form.[55] The problem of how substance is individual is a problem of individuating it only if form as universal is assumed, but for Aristotle substance is a unified self-subsisting thing, not something that must distinguish itself from other beings like it in order to be one.[56] The problem with invoking species form to explain substance is that species form cannot explain what makes a thing individual, which is to say, what makes it be what it is, it can only explain how substances are alike, so it does not even satisfy the requirement of giving knowledge of substance.

Among those who recognise that form is individual, a tendency remains to make substance wholly identical to form on the basis of the view that Aristotle makes form the proper object of definition in *Metaphysics* Z.10. In that chapter, Aristotle distinguishes between the way that flesh is a part of snub (by definition), but bronze is not a part of the circle (incidentally). Frede and Patzig, for example, conclude that since form is the proper object of definition the flesh is excluded from the form. This approach treats material not as a part of what it is to be this substance, but only as part of what it means to be generic natural substance.[57] Including the material aspect of different substances in the essence of the substance strengthens the case that forms are individual, as Whiting argues, and that the natural composite substance is knowable, even as it is caused by an individual form. Aristotle notes the difficulty in distinguishing matter from form in some kinds of substances in Z.11, and proceeds to designate perceptible substances as those that include material in the account of the concrete substance. Whiting argues that this conclusion refers us back to the way that material is included in the definition in the previous chapter as flesh is included in snub. Frede and Patzig

---

[55] See Whiting, 'Form and Individuation in Aristotle'. These arguments seem to conclude from Aristotle's claim that 'substance is predicated of matter' in *Meta*. Z.3 that he is there referring to his own position, rather than notions of both matter and substratum that he immediately proceeds to reject (1029a20-6); see Gill, *Aristotle on Substance*, pp. 25-37.

[56] A. C. Lloyd describes this as the difference between the principle of unity and the principle of individuation. The second begins from the being of the species, while the first begins from the being of the individual in 'Aristotle's Principle of Individuation'. Whiting's distinction between the oneness of species – lacking any contrariety in the account – and the oneness of the individual – by being indivisible in place and time – is a distinction between oneness in account and oneness in being in 'Form and Individuation in Aristotle', p. 372.

[57] Frede and Patzig, 'Sind Formen Allgemein oder Individuell?'

find in Z.10 a distinction between form (concavity) and compounds (snub), but Whiting argues that the distinction is between two kinds of objects or essences – those that include the material necessarily and those that do not, leading her to conclude that material can be included in the form of the composite substance.[58]

Instead of seeing Aristotle's association of substance with essence as an exclusion of material, the argument should be contextualised in light of Aristotle's critique of Plato's Forms. For Plato, the essence of the Form can be otherwise than the Form, which is to say the essence is other than the substance. Aristotle addresses this Platonic problem of the relationship between a substance and its essence in *Metaphysics* VII.6 through what Theodore Scaltsas calls a 'Second Man Argument'.[59] In Third Man Arguments, a third term is needed to unify a first and second term in such a way that continues to gesture toward another term that unifies the new third term and the previous two terms. Aristotle's argument in *Metaphysics* VII.6 is that the introduction of a second term is itself a problem. If substance is distinguished from its essence, and essence is what it is for the substance to be, then what it is for the substance to be is otherwise than the substance.[60] Aristotle argues that if the Form whose content it is to be – Being – is not, then there would be no Forms.[61] Taking the example of the Form that is directly related to being, Aristotle argues if this substance – the Form of Being – could be different from its essence (Being), then nothing would be because the Form that provides for existence could be otherwise than its essence. For this reason, the essence of

---

[58] Jennifer E. Whiting, 'Metasubstance: Critical Notice of Frede-Patzig and Furth', pp. 629-31. Whiting concludes from this view that material individuates form from species form, but this view supposes that species form – and the identity of members of the species – is somehow prior for Aristotle, which would seem to contradict her notion that there can be substantial universals for the sake of definition, whose source is the individual forms that come to be and pass away, pp. 625-31. See also Whiting, 'Form and Individuation in Aristotle', p. 372.

[59] Theodore Scaltsas, 'Aristotle's "Second Man" Argument'. While Scaltsas argues that form cannot be other than substance otherwise the essence of form will be otherwise than the substance, he develops this argument to mean that all the components of substance must exist in the abstract in *Substances and Universals in Aristotle's Metaphysics*. He thus solves Aristotle's ontological problems by making Aristotle's metaphysics into a theory of abstract objects.

[60] David Bostock argues that this reasoning that the essence and substance must be the same is faulty by making the issue a logical rather than ontological problem. Bostock takes the claim that substance is essence to be an argument that form is substance in a Neoplatonic sense, against the view that substance is hylomorphic for Aristotle in Bostock, *Aristotle:* Metaphysics *Books Z and H*, pp. 109-10, 103-7.

[61] *Meta.* 1031b7-20.

the Form must be identical to the substance that is the Form. The problem is not so much that a regress would follow, since Aristotle is elsewhere willing to allow regresses, as for the ontological continuity of natural forms, nor that this is a merely logical or epistemological problem.[62] The problem arises if even one step is taken on the regress away from the substance. Since substance is primary and καθ'αὑτό – it is what it is in virtue of itself – its essence cannot be otherwise than itself.

This problem cannot be solved by the traditional approach of arguing that species form or universal form is most substance, since that would not meet the criteria of separability and thisness that Aristotle establishes for substance at the beginning of *Metaphysics* Z. These criteria lead Aristotle to reject a certain view of substratum in favour of the notion of substance as a 'this' that is the substratum for all other change, which leads Aristotle to propose consideration at the end of Z.3 of how form causes substance to be separable and a this by contrast to the other candidates for substance.[63] By contrast to Plato for whom the Forms are substance and essence, but not a substratum, Aristotle makes substance identical to nature by making what most is – substance – a substratum.[64] Species form cannot be this kind of separable and self-subsisting substratum. Like Plato, Aristotle thinks that that which most is should be defined as form, but that is because form is causal of substance, which is the substratum for everything else that is. Aristotle rejects a certain view of material, the view that makes material bare substratum – in *Metaphysics* VII.3, as will be further considered in the following chapter – not because substratum is not the way to think about substance, but because substance

---

[62] For the view that the problem is the regress, Alan Code, 'On the Origins of Some Aristotelian Theses About Predication', p. 122 and M. Frede and G. Patzig, *Aristoteles Metaphysics* Z, II, pp. 94–5. As Scaltsas argues, 'Proliferation is unwelcome, but in itself is not absurd', p. 128n15. For the view that the problem is a logical or epistemological one, Frank Lewis, 'Plato's Third Man Argument and the "Platonism" of Aristotelianism', p. 164 and Code, 'On the Origins of Some Aristotelian Theses', p. 121.

[63] *Meta.* 1029a26–32. Following Whiting, 'Metasubstance: Critical Notice of Frede-Patzig and Furth', p. 372. Michael J. Loux rejects the need for these criteria for Aristotle in 'Form, Species, and Predication in Metaphysics Z, H, and Θ', distinguishing between the universal Aristotle rejects – genus – and the universal that Aristotle accepts as a candidate for substance – species. Michael J. Woods argues for the separability and thisness of species form in 'Form, Species and Predication in Aristotle'. Woods's argument depends on same (ὁμοειδής) in *Metaphysics* VII.4 meaning the very same form rather than similar form contra Frede and Patzig.

[64] *Meta.* 1031b16–17.

is not a bare substratum, but what is self-subsisting, and matter is not that which remains when all else is stripped away. Aristotelian form does not just define substance, but wholly causes and characterises it, while Platonic Form does not. On the Platonic notion that Form is substance because Form itself is self-subsisting, Form would be causal only of itself as substance and only incidentally of hylomorphic sensible beings, which would not be considered substances or substrata, contrary to Aristotle's many claims that substance is substratum and his commitment to the phenomena, whereby the composite substances appear to be most what is.[65]

Not only does Platonic Form not fully characterise that which it causes, it is not even wholly self-predicating of itself. Plato can open that distance between Form and nature, while Aristotle cannot because, as we learn in Aristotle's critique of the view that there can be Forms of non-substantial things in *Metaphysics* 990b22–991a2, Aristotle maintains that form wholly characterises and governs that of which it is the form, while Plato, or at least, the Neoplatonic reading of Plato, does not. Aristotle argues that for Plato, the nature of the Form appears contrary to the nature of the content of it as in the case of the Forms of attributes. Aristotle argues that on Plato's own terms, the content of Forms of attributes would be inconsistent with the nature of these Forms, so Plato should not permit Forms of attributes. If he did, then the Form of the Double, for example, would be accidentally eternal, since the Form is eternal, but we know that doubles need not be eternal, so the Form

---

[65] Some scholars base their argument for the priority of species form entirely on *Metaphysics* Z and so are less inclined to consider natural hylomorphic beings substance. These arguments that depend on a developmental argument are less convincing than a reading that aims to account for the whole of the text. Aristotle speaks of natural things as substances throughout *Metaphysics* VIII and in *Phys.* 192b33, *Meta.* 1028a20–30 and 1028b26–31. At *Meta.* 1032b14–15, Aristotle calls essence the substance without matter, which would be a strange configuration if he meant that essence is substance that is predicated of matter rather than that substance is essence and matter. Similarly, at *Meta.* 1033a30–1, Aristotle says that 'to make a "this" is to make a "this" out of the general substratum', suggesting that the 'this' that is the substance is not solely the form but the composite substance. He discusses the 'matter of such substances' at 1037a10. He refers to 'concrete substance (ἡ σύνολος λέγεται οὐσία)', a number of times in this chapter at 1037a30–2. Admittedly, the distinction between substance and the concrete thing at 1035b14–21 poses a problem, but Aristotle concludes this chapter by considering whether the soul is identical to the individual animal in a way that suggests the distance between the essence and the particular is not so great. Aristotle further addresses the question at *Meta.* XII.5, when he argues that universal causes do not exist, but that individuals are the cause of individuals, because there is no universal man, only individuals, who come from their fathers at 1071a19–24.

would be internally inconsistent.[66] Avoiding this problem requires forms to be only of those that are themselves what they are in virtue of themselves, but for Plato these criteria only hold for Forms. If things exist and are then characterised by form, the result is what Scaltsas calls a 'nature-feature' problem, where the nature of a thing is distinct from the features that characterise it. This view is related to what Witt calls an 'object-property' view, where the essence characterises, but does not cause that of which it is the property in a way that makes form predicated of substance, rather than causal of it.[67] The view that form is substance and causal of substance because form is substance rather than the composite, which is not caused by form but by substance, which is predicated of matter following a reading of Z.3, does little to explain the central questions of *Metaphysics* Z.7-9, which focus on natural substance.[68]

Substance can differ from other substances that have the same form, sexual difference and inherited traits being examples of how. And these traits can be knowable, we learn in the biological works, to the extent that sense perception is a source of knowledge. As Witt explains, essences differ by being causal of different substances because form individuates, a position that is compatible with the view that composites differ by being composed of different material.[69] The individual nature of form as soul is supported by Aristotle's argument in *Metaphysics* E.1 and in *De Anima* II.1 that the soul as the actuality of the

---

[66] Julia Annas's objection to Aristotle's argument against Forms of attributes is that Aristotle refers to the second Form as accidental in a way that is not found in Plato's account of how the second Form follows from the first and that Aristotle's critique depends on a special case, that of the eternal in Julia Annas, 'Aristotle on Substance, Accident and Plato's Forms'. The second objection seems to be that the eternal is a problem for what it means to be Form, but this is precisely Aristotle's point – that certain things must be characterised of the essence of the Form that are contradicted by the content of the Form and thus that the essence of what it is to be Form is at odds with what it is to be this Form. Pieter d'Hoine argues that Aristotle is not referring to an alternative kind of participation but to participation in a Form qua attribute of another Form such that the double as a substance (on Plato's terms) is participating in the eternal. D'Hoine argues that Aristotle uses Platonic examples in order to argue that even on Plato's terms Forms should only be of what Aristotle calls substances in d'Hoine, 'Aristotle's Criticisms of Non-Substance Forms and Its Interpretation by the Neoplatonic Commentators', pp. 271-2.

[67] Witt argues against the object-property view that makes form predicated of substance that form is predicated of material in *Substance and Essence in Aristotle*, p. 125. This view of predication would seem to entail the same problem of making material the separable substance of which form is predicated instead of the potential that form actualises.

[68] See Code, 'Aporematic Approach to Primary Being', p. 18.

[69] Ibid. pp. 176-8.

body is inseparable from it. Aristotle recognises that some parts of the soul are separable because they are not an actuality of the body – the rational part of the soul. But this separability does not make the soul universal. It remains the cause of this individual. Aristotle maintains that the soul is the essence *because* it is that 'by which primarily we live, perceive, and think'.[70] Aristotle concludes that the soul is the actuality of that which is potentially this being. This view of soul requires soul to be specific to each individual body of which it is the actuality.

Form actualises what is potential, which is material that is both inseparable from form in the generated substance and retaining its own character as elemental material. Substance as form is form actualising material, which is why the form must continue to unite and organise the natural substance. The potentiality is not used up in generation, because material remains capable of its material capacities insofar as the general substance can decay. Actual knowledge is of the individual essence, while the potential knowledge is of the universal, what the essences have in common. Aristotle ends *Metaphysics* VII with a return to this point, asking in VII.17, '[W]hy is this individual thing, or this body in this state, a man?' It is in answer to this question that Aristotle writes, 'Therefore what we seek is the cause, i.e. the form, by reason of which the matter is some definite thing; and this is the substance of the thing.'[71] The cause he pursues in what follows is explained in terms of actuality, which is separate and individual in each substance, and both form and matter.

Form as the cause of individual perceptible substances is multiply related to material: it works through material as semen in generation, it requires functional matter to persist in fulfilling its end, and it works on individual material in the female in order to generate. In order for Aristotle to maintain the position that the substance is essence, the semen must be the power of the soul as semen, rather than have this power as something distinct from what it is. As Roberto Lo Presti notes, this position is often ascribed to the Hippocratics, but it also characterises Aristotle's account of how the female matter receives the soul 'in the form of an internal principle of organization' that the semen by virtue of the material capacities that comprise it sets in motion.[72]

Neoplatonists who argue for a distinction between Form as such and enmattered form or between Form and nature deny that Form as such is working through enmattered form. Their account continues to influence readings of

---

[70] *De An.* 414a4–7, 13–14.
[71] *Meta.* 1041b7–9.
[72] Roberto Lo Presti, 'Informing Matter and Enmattered Forms: Aristotle and Galen on the "Power" of the Seed', pp. 934, 937.

form in Aristotle.[73] Efforts to make Aristotle's position consistent with Plato's ignore the different reasons Plato and Aristotle invoke form. Aristotle uses form to explain how substantial change can occur and substance be what most is: change occurs between contraries of the privation of form and form, and this form wholly characterises what has come into being. Plato appeals to Form to explain the co-existence of contraries within the same entity.[74] That something can be attributed to a thing can be explained by the thing's participation in the Form of that attribute. Participation in Forms can explain how Socrates can be both great and small because neither fully characterise Socrates, so opposites can inhere in the same entity without being of one another. Aristotle's concepts of form and substratum allow him to distinguish between generation, wherein the substratum becomes what it most is through the generation of form, and privation and alteration, wherein a pre-existing substratum as substance takes on one form while retaining its nature. All change for Plato, by contrast, is extrasubstantial, not only because Forms are the only substances and they do not change, but also because Forms do not wholly characterise the nature of what participates in them, which might be said to lack a nature of the Aristotelian kind. All attribution is based on participation that does not fully characterise that which participates because if it did then opposites would belong to one another. That which participates is somehow always otherwise than that in which it participates, a problem which begins with attributes, but continues in the problem of distinguishing between a substance and its nature. This approach opens a distance between what a thing is and its characteristics, between properties and objects, features and natures in a way that continues to influence readings of Aristotle.

Responses to Aristotle's critique of Forms of attributes insist on the distance between the transcendent Form and a thing or its nature, resulting in the problem where the essence of Form as Form contradicts the essence the Form predicates. G. E. L. Owen responds with a strategy developed by Proclus whereby a Platonic Form can be defined on two levels – what Owen calls the P-distinction.[75] On the first level of the Form qua Form, the Form is defined in

---

[73] Michael Frede argues that form is substance because knowability is the criterion for substance, which leads Frede to the view that Aristotle establishes individual substantial forms as primary entities in Michael Frede, 'Substance in Aristotle's Metaphysics'.

[74] This understanding of the purpose of Plato's Forms is so described by Scaltsas, 'Aristotle's "Second Man" Argument', pp. 118–21 and d'Hoine, 'Aristotle's Criticism of Non-Substance Forms and Its Interpretation by the Neoplatonic Commentators', p. 265, and seems uncontroversial.

[75] G. E. L. Owen, 'Dialectic and Eristic in the Treatment of the Forms'.

virtue of its status as Form, what Owen calls A-predicates. On the second level, the Form is defined as that which it characterises, B-predicates. Owen argues that these two levels can be held apart so that what is true of A-predicates has no bearing upon what is true of B-predicates. This approach sidesteps the problem of the essence being other than that of which it is the essence only to the extent that not the Form but that nature that is its content is the essence of that which participates in it.

Proclus argues for four stages of ontological existence in order to explain how Forms can be fundamentally causal without themselves having a relation to sensible beings. Forms, reason-principles in soul, reason-principles in nature, and sensible things or bodies make up the causal chain. While Forms are absolutely transcendent, they cause sensible beings, because each thing is what it is by reflecting the next level, so Forms can be known at each level by their reflection, but never on their own terms, since they do not inhere in enmattered beings (except perhaps through being implanted in the soul by God). In generation, Nature, at an ontological remove from semen, actualises the reason-principles in semen because Nature has reason-principles in actuality, even though nature itself does not have reason or imagination, which are found in Soul. Proclus describes Nature as a carpenter who enters the wood and shapes it from within, in a way that makes even an internal source of movement external and imposing, rather than wholly integrated in what it shapes as the shaping shape itself. Because bodies require an 'indwelling cause' that also has reason-principles whereby it can guide those principles, Nature has to have the reason-principles to guide it. Insofar as Nature needs these principles, it is without reason and therefore not a genuine cause, which must be intelligent and divine. The body that nature causes from within is the image of these principles of Nature, which are themselves the images of principles in the soul, which are themselves images of the Forms. Because Forms cause immaterially through reflection and not bodily, Forms can be common to and present in the many things that participate in them.[76]

Proclus describes the Forms as patterns, but causal patterns rather than sterile and lifeless ones. He argues that the Forms cannot be causing generation through the things in which they are found, referencing the Aristotelian phrase, 'for man begets man', because then the cause would not be 'absolutely eternal', but 'subject to motion and change', and so not a reliable cause. Aristotelians are more willing to allow that the forms are persisting in the sense that an individual

---

[76] Proclus, *Proclus' Commentary on Plato's* Parmenides, 891.32; 897.37–898.1; 949.13–18; 950.5–7; 792.5–19; 794.3–20; 795.26–7; 797.19–32; 844.1–5; 883.1–6, 20–2.

form continues to be at work in a natural substance so that the universal remains without having to have a distinct ontological status from the individual form from which it draws its existence. In order to preserve the eternal qua unchanging character of forms, Proclus explains that Forms are causal of material things with which they have 'nothing in common', by causing through reflection, so while they remain unmixed with the lower levels of existence, they do commune with their participants 'to the extent that they will illuminate them from their own essence'. He addresses the problem of how Form causes individuals without having to argue that material is the individuating cause because his Forms cause through reflection. The Forms can govern the material world and be responsible for it on this account while still being transcendentally superior because the four levels of being ensure that there is no relationship of the Form to the generated thing. Sensible things are homonymous with intelligible ones because the sensible are only a reflection of what is higher up the scale of being, and so the content of the Form exists through the chain of the Form (the series of levels that participate in the Form by reflecting it), where essence of Form is in no way shared with the lower levels of being. Plato through Proclus has a problem of the knowability of substance, insofar as Form is substance, because they are not in us so our faculty of knowledge fails to access them because they are fully transcendent.[77] We can only ever have knowledge of what is reflected in the reason-principles of the soul.

Proclus's representative Neoplatonist interpretation of Plato provides the background for why commentators on Aristotle insist on a distinction between form and material that forbids the operation of form through and in material. But Aristotle rejects some of the key investments that Proclus finds in Plato. For one, Aristotle argues for the primacy of individual substance, and forms are most this substance. For this position to hold, forms cannot exist on another ontological level from those natural substances that they cause. Natural forms do pre-exist, but immanently, in another being of the same kind, the father. The natural form of the father is causal in such a way that makes the new substance become self-governing. Natural forms are causal in what they cause. Forms do not cause by reflection, but by being what they are, by actualising. Causing by reflection would introduce a separate power between being what they are and the form itself.

The distinctions between form, agent, tool that Aristotle relies on to explain generation reflect the ontological levels that Proclus finds in Plato,

---

[77] Proclus, *On Parm.* 910, 16–19; 884, 1–7; 956, 21; 889, 21–34; 890, 4–12; 921, 1–21; 923, 15–27; 939, 17–18; 904, 4–11; 949, 9–12.

but Aristotle does not employ the analogy in order to maintain this ontological distance. The immanence of Aristotle's form and the form as the cause of substance keep Aristotle from needing these distinctions. Aristotle is critical of the view that requires them in order to make sense of form's work (Owen's P-distinction), because for him, the form is not otherwise than the natural substance, but the potential knowledge of the individual substance that is actually knowable and the active ordering that makes the substance continue to be what it is. Aristotle's account of perceptible substances even offers room for material to be a part of the account of the substance – part of what it is for the substance to be what it is. If form is the father, the agent is the semen and the tool is the heat and the air, then it seems like this structure does not map onto the ontological distinctions of Proclus because each of these are material composites.

A useful contrast to Proclus's attempts to harmonise Platonic and Aristotelian metaphysics is Albertus Magnus. Albert the Great offers what Michael W. Tkacz calls a 'de-Platonized conception of Aristotelian form'.[78] Albert argues that Aristotle rejects the view of Ptolemy that intelligible forms must be distinct from matter and argues instead that forms exist and come to be in matter that is potentially formed. He likewise rejects the position of the Platonists that natural species are immutable ideas that are separate from the particulars, which for the Platonists are only reflections of the ideas in the sensible world. For both Ptolemy and the Platonists and Neoplatonists, sensible subjects cannot be the source of scientific knowledge. But Albert argues against the Neoplatonist view that form is knowable through a mathematical reductionism, associated with the Oxford Platonists. Albert supports the notion that observation of the physical world and of sensible beings coupled with theoretical principles allows us on an Aristotelian account to determine the relation of form to function in the world.

The argument that form has to be intellectible in order to be knowable or in order to be held in common loses its force when form is understood as individual and the definition includes the material. My argument is not on the side of the materialists, that material alone, without a purpose toward which it is organised, can account for the being of things. My argument is that natural form is acting in and through material all the way down to that

---

[78] Michael W. Tkacz, 'Albertus Magnus and Recovery of Aristotelian Form', pp. 739, 753–4. For the critique of the view that intelligible form does not exist in sensible things see Albertus Magnus, *Physica* 1, tr. 1, c. 1–2 in ed. Colon. 4/1:3–5. For the critique of the view that form is strictly mathematical see *Metaphysica* 1, tr. 1, c. 1 in ed. Colon., 16/1:1.24.

which is the work of animation in semen. Natural form does not work without matter. While matter does not coincidentally move toward the end, it is still through matter that form moves to the end, and it is materially that form does so. The material capacities of form in semen contribute to the work of semen. The form of semen in this sense is not separable from the material that makes the work possible. This intertwining of form and matter is what I mean by a Möbius strip notion of Aristotelian hylomorphism. If form is wholly other than material, as the two-sex model would suggest, then form cannot act on matter and matter cannot be acted upon by form. This problem is the one Solmsen, Balme, Tress and Coles each articulate as the need to find a material basis for soul while resisting a reduction of soul to matter. The problem is with supposing that natural form, which exists as enmattered in natural substance, works as otherwise than in and through material.

Conclusion

A number of assumptions about how form and material act are at work in these debates. Commentators argue over whether a particular kind of material that seems to transcend the bodily is responsible for causing life and whether material can in any way be responsible. Form is understood as a power of life that exists distinctly from material rather than the actuality of a potential. References to material as part of the causal apparatus of generation lead to arguments that these materials are not in fact material but more like form, precisely because they have animating power. Explanations of life that depend on the movement of semen are questioned because they are not sufficiently separate from material, assuming that for form to do this work it must be a separable principle. A role for material is allowed in some accounts only if material is a tool of form. The tool is viewed as some separate extension of an existing entity that uses the tool rather than a way of manifesting that of which it is the tool. The restricted material view, what is often called the anti-reductionist view, is motivated by a theoretical commitment to distinguish between the active and directed work of form and the passive and merely necessary work of matter. These accounts tend to consider that form must work separately and observably distinctly from material rather than through and within material. These views suggest that the material the natural substances come from are the same materials that constitute them, and so take claims that the father contributes no material to mean that there is no material significance or power in semen that enables it to animate the menses. They maintain that material cannot be a part of this enlivening work because then no distinction could be drawn between living and non-living things, again a position which supposes that form pre-exists its involvement with material in generation rather than that the work of form is observable in living. These arguments routinely reject the material explanation of semen on the basis of theoretical commitments to a view of form that works separately from rather than through material as the actuality of it.

These readings depend on understanding form as species form. As species form, form is most ontologically real outside of individual substance, which explains how it can be thought as an intellectual rather than material force in natural substance. Strong cases have been made in the last forty years that form is individual. In order for it to be causal of individual substance, it must be individual. Definition is potential knowledge of what is held in common between similar forms, but actual knowledge is of individuals. Further cases have been made to support the view that the definition includes both form and matter in natural substances, because of the indivisibility. The tendency to find in Aristotle a distance between the work of form that he describes as enmattered and the work of form itself as not enmattered can be traced to Neoplatonic programmes to protect the eternity of Forms from their individual instantiations. These concerns do not motivate Aristotle, I maintain, and his craft analogy, while it could be mapped onto the Neoplatonic ontological distinctions, is not employed by Aristotle toward that end.

The case of semen as both the site of form and as explainable in material terms is indicative of the interdependence of these causes. Commentators who are sympathetic to the work of the material of semen in generation largely focus on the movement of semen that the semen takes on from the parent that works on the material to make the work of generation immanent to the joining of semen to menses and not externally motivated by some formal agent. Notably material is still portrayed as inert, even bordering on unnatural, even though the menses is a residue of natural processes in a natural being. Both menses and semen are explained as developed through a kind of elemental power – heat. Aristotle's material is not inert, even elemental material has its own power, which it continues to maintain when it is taken up into natural substance. The power of elements shows why Aristotle rejects prime matter, a concept that in many ways still structures disputes over Aristotle's biology, which is the focus of the next chapter.

3
Aristotle on Material

Beyond Prime Matter

One persistent obstacle to the project of resuscitating matter in Aristotle's thought is the widely held view that Aristotle maintains a conception of prime matter. If such a concept is at work in Aristotle, whether wittingly or not, then it would seem that matter as such has no power or character of its own. If matter has no power of its own, then its appearance in the world is entirely dependent on form. Material itself would bring no contribution, except the occasion for form to manifest itself in the world. Natural substances would owe their being to form with the material making no positive contribution except as the status of what gets formed. The hylomorphism of substance would be only weakly interdependent because material's role would only be to exist, compliantly, for form to do its work. Form's dependence on material would not be due to the capacities material brings to the generation or persistence of substance, but only as pure capacity itself. Finally, and most importantly for my account, if Aristotle has a conception of prime matter then he has an account in which form and matter are wholly separable, an account which follows the craft model of nature. Form would be best captured as the efficient cause, moving in and upon completely uncharacterised powerless stuff to make it into substance.

If Aristotle's account of natural generation is one of the craft model, then Irigaray is right to say that material is merely for the sake of form. Prime matter is that which is incapable of change on its own and has no meaning of its own. Prime matter is the material Irigaray describes as the forgotten outside and what Bianchi calls the pure potential that is also robbed of its own potential. Prime matter is the matter that Butler is concerned in Irigaray's reading produces material as bare natural givenness.[1] The craft model of generation

---

[1] Luce Irigaray, 'How to Conceive (of) a Girl', in *Speculum of the Other Woman*, p. 164; Emanuela Bianchi, *The Feminine Symptom: Aleatory Matter in the Aristotelian Cosmos*, pp. 183–222; Judith Butler, *Bodies That Matter*, p. 5.

depends on a conception of prime matter because it depends on a complete separation of form from matter. The craft model then construes matter as not form even while it supposes that material and form offer distinct contributions. If Aristotle's metaphysics includes a concept of prime matter, then he construes material as an occasion for form to work, having no positive power of its own.

If Aristotle does in fact have an account of prime matter, then my argument that form and matter are interdependent and that form in semen depends on material powers to animate the menses would not make sense of Aristotle's account of form and matter in natural substances and in natural generation. My argument would not make sense because material as prime matter would not have its own power. My attempt to present a more robust role for material would falter because form would have to be understood as wholly separate from material and if dependent on it at all, only dependent to manifest itself, rather than to contribute to its work. The argument of this book, then, that matter and form should be understood as interdependent, along the lines of a Möbius strip model, depends on debunking the view that Aristotle has a conception of prime matter. That is the project of this chapter.

This chapter considers the key passages in Aristotle's corpus that contribute to a view of matter as what is fundamentally without its own character alongside recent literature that argues against the existence of prime matter in Aristotle and against taking over too thoroughly the analogy from natural substances to artifice. These passages point to a view of material that shows material to have its own character, form not material to be the substratum that is substance, and the material of both καταμήνια, menses, and γόνη, semen, to be transformed in generation. This reading supports the notion that semen works through material while still allowing it to function as the formal rather than material cause.

The previous chapter concludes with an argument for how to understand form in Aristotle's work as individual and as acting in a way that finds it interwoven with material. This chapter will show how disputes over prime matter reveal a host of additional issues in Aristotle's metaphysics into which his biology gives us insight: whether material or form is the substratum in natural substance, since material seems to be the substratum in artefacts; whether material and substratum are pure potentiality without character; whether the elements have a common matter that is its own distinct ingredient from the elemental forces that constitute elements; whether generative material is equivalent to constitutive material in natural substances as it seems to be in artefacts; the extent to which material has its own character in natural and artificial production; whether the substratum is always a composite; whether the actualising of a potential is a change from a privative actuality to a positive actuality and a positive potentiality to a privative potentiality or from a positive actuality to a positive actuality; whether first matter is a technical term or reference to the closest, proximate matter; whether something insubstantial must persist through substantial change; whether all of Aristotle's references to matter are references to his view or to the way that people speak of material. This examination of prime matter in conjunction with the specifics of Aristotle's biology will show that the substratum in natural substances always has a distinct character, that neither material nor the substratum in natural things is ever pure potentiality, that the common matter of elements is the elemental force they share in generation or destruction, that generative matter is in one way distinct from constitutive matter in natural substances and in another way not, that material is both fully characterised by the substance it constitutes and that material maintains its own generic material nature, that the substratum of natural substances is always composite, that the actualising of a potential is a move from positive to positive, that first matter is not a technical term, but a referent to the closest matter in some passages and to a shared material in a particular situation in other contexts, and that Aristotle sometimes refers to the ways people speak of material in order to take up the truth of it and to consider the problems with it.

Aristotle's material has its own character and this character is not that of Cartesian extended things, but more robust, 'powerful, rather than . . . stuffy', as Kosman puts it.[2] Robust material is the other side of the hylomorphism whose form in natural generation and natural persisting substance is not some hovering power in which material must participate, as the Neoplatonic early readers of Aristotle suggest. Hugh R. King argues that in their effort to make Aristotle compatible with Plato, early interpreters of Aristotle transform Aristotle's underlying material into Plato's χῶρα. These thinkers did not so much argue for this view from the text as suppose it followed from Aristotle's account of matter and form.[3] King argues that this view contributed to the tradition of reading matter in Aristotle as inert and form, not as the organisation and activity of matter, but as something disconnected from matter and to and in which matter must aspire and participate.

---

[2] L. Aryeh Kosman, 'Animals and Other Beings in Aristotle', p. 382.
[3] Hugh R. King, 'Aristotle without Prima Materia', pp. 388–9. King quotes Joachim, 'Speaking for the Tradition, Joachim says that what happens in generation is "the transformation of a permanent substratum whereby it drops one form and takes another".' Joachim, *Aristotle on Coming-to-Be and Passing-Away*, pp. 92–3, 97; King, p. 374. See Mary Louise Gill, *Aristotle on Substance: The Paradox of Unity*, pp. 27–9.

Aristotle sometimes speaks of material as a first matter when he seems to mean the closest material, what King calls the 'proximate matter', which is akin to the last matter, that which makes it appropriate for the substance it becomes.[4] One passage that establishes some key principles of Aristotle's account of material is in *Metaphysics* V.4, when Aristotle defines nature. His fourth definition of nature is of first or prime matter. Within this account, he distinguishes between the way that something is the matter of artifice and of natural substances:

> Nature is the primary matter of which any non-natural object consists or out of which it is made (ἐξ οὗ πρώτου ἢ ἔστιν ἢ γίγνεταί τι τῶν φύσει ὄντων), which cannot be modified or changed from its own potency, as e.g. bronze is said to be the nature of a statue and of bronze utensils, and wood the nature of wooden things; and so in all other cases; for when a product is made out of these materials, the first matter (τῆς πρώτης ὕλης) is preserved throughout. In this way people call the elements of natural objects also their nature, some naming fire, others earth, others water, others something else of this sort, and some naming more than one of these, and others all of them.[5]

The first matter of artificial things is that from which they are made, which if it is changed to be something other than that material would no longer be potential for this artefact. In these artefacts, that from which it comes-to-be remains as this material throughout the artefact's existence. Aristotle calls attention to the parallel between this kind of material and the elements of natural objects which constitute natural objects and seem to remain as those elements but further formed in natural substances. He is not stating his view that the elements are prime material, though that might seem to be the case, but rather explaining why people say that they are.

This chapter is organised around the passages where Aristotle discusses the role of material in his corpus, *Physics* I.7, *Metaphysics* Z.3, *De Caelo* and *On Generation and Corruption*. From these passages, material will show itself to have its own nature as elemental – it is not a merely Platonic receptacle for form – and to be transformed in natural generation in a different way than in artificial generation.

Physics I.7: Change and the Substratum

Physics I.7 seems like a good place to begin because in this chapter Aristotle posits a ὑποκειμένων, a substratum, to respond to the sophistic argument that

---

[4] King, 'Aristotle without Prima Materia', pp. 372–3.
[5] *Meta.* 1014b26–35.

generation is impossible because something cannot come from nothing. Aristotle introduces the problem with a consideration of how we speak of accidental change in a substance, as when we speak of a man becoming musical.[6] We speak of that from which change comes in both simple ways – the man or the unmusical – and complex ways – as the unmusical man.[7] The two ways of speaking of that from which something comes to be point to two ways that from which something comes to be is at work in what becomes. In the one sense – the sense of man – it survives through the change. In another sense – the sense of unmusical – it does not survive. The question facing Aristotle is what survives and what does not. He concludes from these different ways of speaking:

> These distinctions drawn, one can gather from surveying the various cases of becoming in the way we are describing that there must always be an underlying something (ὑποκειμένων), namely that which becomes, and that this, though always (εἰ καὶ) one numerically, in form at least is not one. (By 'in form' I mean the same as 'in account'.) For to be a man is not the same as to be unmusical. One part survives, the other does not: what is not an opposite survives (for the man survives), but not-musical or unmusical does not survive, nor does the compound of the two, namely the unmusical man.[8]

The case of change that Aristotle uses here as a paradigmatic one is accidental change, and what underlies that accidental change is the natural substance that has its own being. One argument in support of prime matter is that Aristotle is arguing that this structure of change of the accident in the substance which underlies is analogous to the change in substance, where material would remain the same in that change just as substance does. This analogy is why the partisans of prime matter insist that the material of which a substance is made remains the same in generation, and why others argue that material is always transformed in generation.[9] The argument in support of prime matter

---

[6] Barrington Jones argues in 'Aristotle's Introduction of Matter' that Aristotle argues from how we speak, what Christopher P. Long will come to call 'legomenology' in *Aristotle on the Nature of Truth*. Pace Alan Code, 'The Persistence of Aristotelian Matter'.
[7] *Phys.* 189b33–190a4.
[8] *Phys.* 190a12–21.
[9] Jones argues that even in artifice the material is transformed, but this need not be the case to say that generation transforms material because artificial generation is akin to the accidental change. Jones, 'Aristotle's Introduction of Matter', pp. 485–6. Theodore Scaltsas maintains that in generation, everything – both the original matter and the original kind of material – is transformed, with the implication that material must then, like the form and the composite, be an abstraction in *Substance and Universals in Aristotle's* Metaphysics, pp. 16–18.

would require that substance be the same in all accidental change, as prime matter is, which is clearly not the case. It also relies on reading 'εἰ καὶ' as 'though always', rather than as Cohen suggests, καὶ εἰ, 'if indeed (though I don't think it is)', which would suggest that even if it is one in number, it is not one in form to show that Aristotle is not affirming a single substratum for each change.[10] After returning to the various ways we speak of what becomes and what it becomes, Aristotle notes,

> Things are said to come to be in different ways. In some cases we do not use the expression 'come to be', but 'come to be so-and-so'. Only substances are said to come to be without qualification.[11]

Both come to be from something underlying, but this sense of underlying differs in accidental change and substantial change, for as Aristotle explains, when an attribute comes to be 'a subject is always presupposed', and the reason is that 'substance alone is not predicated of another subject, but everything else of substance'.[12] That which underlies the accidental change is a substance, which exists on its own accord and is not predicated in something else, while accidents must always be predicated of something else. Aristotle does argue that substances come to be from some underlying thing, but the examples of how substances come to be from an underlying thing are not that which substances are composed of but that from which they come to be – 'for instance, animals and plants from seed'.[13]

All the changes are complex because every change always includes something which comes to be and something which becomes that, both as the opposite of what it comes to be and as the ὑποκειμένων. Aristotle concludes that 'everything comes to be from both subject and form',[14] which are three principles in every change – the contraries of form and the ὑποκειμένων.[15]

The focus of this argument has been that what comes to be comes to be from something, what Eric Lewis calls the 'ἐκ relation'.[16] Aristotle has now established two ways that something can come to be from an underlying thing: in one way, it comes to be from something which continues to be throughout

---

[10] Sheldon Cohen, 'Aristotle's Doctrine of the Material Substrate', p. 182.
[11] *Phys.* 190a31-2.
[12] *Phys.* 190a33-6.
[13] *Phys.* 190b4.
[14] *Phys.* 190b20.
[15] *Phys.* 190b29-191a7.
[16] Eric Lewis, 'Introduction', pp. 19-21, 35-9, 54-6.

the change; in another way, it comes to be from that which precedes it, out of which it comes to be. Both kinds of change are in a sense from substance: the first is accidental change that inheres in substance and the second is from a prior composite that is transformed in generation. The first change is accidental change and the second is substantial generation. These two changes share the same paradigm because they both involve three principles of change and in both changes, the material out of which something comes into being is a composite with its own being (and this paradigm will hold even for the simplest bodies, the elements).[17]

Partisans of prime matter argue that Aristotle's closing analogy in this chapter points to the ὑποκειμένων being the same in both kinds of change:

> The underlying nature can be known by analogy. For as the bronze is to the statue, the wood to the bed, or the matter and the formless before receiving form to anything which has form, so is the underlying nature to substance, i.e. the 'this' or existent.[18]

The ὑποκειμένων is related to what becomes as something present in it but altered. The human being is formed from seed and the bed from wood. Bronze is what the statue comes from as the wood is what the bed comes from and matter what anything which has form comes from. In this way, the substance comes from the underlying nature. That coming from is the ἐκ relation in the analogy, rather than that which constitutes the 'this'. The material in both the artefact and in the natural thing is a composite that exists before it becomes what is generated, but the natural thing comes from seed in a different way than the box comes from wood because the seed is transformed in the generation of natural substance.

Advocates of prime matter argue that the relation described here is transitively explaining ὑποκειμένων as prime matter: insofar as bronze is to the statue, so must some more basic existing thing be to the substance, thereby arguing that ὑποκειμένων refers to another distinct and most basic ingredient that must remain the same through change, rather than any particular thing acting as material in a particular change. As Gill argues, nothing in this chapter points to an ultimate substratum in need of characterisation nor a ὑποκειμένων lacking its own identity. The chapter aims to explain how change involves a third principle, which in every case has its own character.[19]

---

[17] Kosman, 'Animals and Other Beings in Aristotle', pp. 362–3; Gill, *Aristotle on Substance*, p. 104.
[18] *Phys.* 191a9–12.
[19] Gill, *Aristotle on Substance*, pp. 106–7.

Aristotle does not mention material until the end of the passage where its relation to form is associated analogically with ὑποκειμένων's relation to substance, so material does not seem to be the focus of the chapter. The way that the matter that preexists what comes to be 'a this' remains in the 'this' is not elaborated here, and it appears different in accidental change than substantial change. The wood of the threshold, to use Kosman's example, is what it is by being in a certain place, which makes its unity with its form precarious and thus analogous to the unity of an accident in a substance. In the accidental cases, the ὑποκειμένων has its own being and that which changes is only precariously unified with it.[20] The lack of unity permits what has come to be the ὑποκειμένων in the new thing to persist. This lack of unity makes the change an alteration rather than a generation. By contrast, in the generation of natural substances from seed, the unity is central, and the unity means that the ὑποκειμένων is fundamentally transformed, rather than persisting, in generation. Substances come to be from something that underlies in the sense of that which came before, but the process of generation fully transforms that from which it generates.

Commentators argue that the seed Aristotle says human beings come from could be both male or female in *Generation of Animals* so Aristotle is likely referring here to the seed as embryo, because that is directly that from which the human being comes to be.[21] The reference to seed here is often assumed to indicate semen, but that reading would make a thing come to be from its form, instead of the form being that which causes it to be substance. Form is a cause, but not what a thing comes to be from, as Aristotle speaks of what underlies in this context. Substance actualised as form can be what underlies, as Aristotle's consideration of substance as substratum in *Metaphysics* Z.3 will show. The wood is in a sense substance of the tree that has its form as tree, which makes it appropriate for being a box. Its capacity to underlie the box – to be what the box comes from and what constitutes the box – is due to the natural substance that it is. The embryo, like the wood, is formed and appropriate for becoming the product. Yet unlike the wood, the embryo is not the material that continues to constitute the box when actualised. Aristotle does refer to καταμήνια as seed

---

[20] Kosman, 'Animals and Other Beings in Aristotle', pp. 368–70.
[21] Gill argues that it is embryo, *Aristotle on Substance*, p. 229, following Jones, 'Aristotle's Introduction of Matter', pp. 489–92. Code argues that man comes to be from seed as efficient cause so that this passage is irrelevant and that human beings are made of flesh since the primary nutrient is converted into flesh upon fertilisation in 'The Persistence of Aristotelian Matter', p. 364. This view depends on seeing seed in this passage as both efficient cause and as material, rather than embryo.

and καταμήνια is concocted blood, but that understanding of seed would not explain how natural substances come from a ὑποκειμένων that is distinct from the ὑποκειμένων that constitutes it.

This shift between the material that comes to be and the material that constitutes a natural thing explains why Aristotle describes matter in *Physics* I.9 as accidentally what is not. It is not what is not in its nature, which is to say it is not not-being, but it is only what is not relative to the substance that forms out of it. By contrast, the privation of form is not being in its own nature.[22] Prime matter supporters who refer to Aristotle's claim in the next paragraph that matter desires form as female desires male and the ugly beautiful ignore the implication of this earlier passage in interpreting that passage, since this first claim establishes that matter's not-being is akin to the man not being musical not to the privation of musical.[23] Moreover, that passage explains how the contraries would seem apparently to desire their own negation. Aristotle explains that it is not that the contraries desire the negation, but that what underlies pursues the change, as the man desires to be musical, rather than the unmusical desires to be musical. That is why he clarifies the claim by saying that it is the ugly and female not in itself but accidentally that desires the beautiful and the male – that being which is ugly desires to be beautiful as that being which is female desires to be male, so that being from which the substance comes to be desires the form that will make it substance.[24]

King argues that prime matter supporters make two metaphysical assumptions about material that lead them to conclusions not supported by Aristotle's text. First, they make out of a principle or role an actually existing thing.[25] Whenever Aristotle refers to some common matter with reference to some thing that plays the role of material between two things capable of changing into one another, they argue that Aristotle is positing a common material. Second, they define material itself as lack of form, which allows them to argue that it must be entirely without character. Having denied any character to whatever is not form, they dismiss as candidate for material whatever has its own character, like the elements or a natural substance from which another substance may generate. Even those who oppose the existence

---

[22] *Phys.* 192a4–6.
[23] A reading that can be traced back to Eduard Zeller, *Die Philosophie der Griechen in ihrer geschichtlichen Entwicklung*, 1844–52.
[24] *Phys.* 192a23–4.
[25] King, 'Aristotle without Prima Materia', p. 387.

of prime matter can be susceptible to these assumptions, as when by insisting on material as a principle, they describe it as a relation where material is what is without its own character.[26] Gill argues for the unique and specific nature of the elements themselves which remain at work in natural substances even as they are transformed into something more complex within the composite. Matter's capacities as generic matter – the matter that constitutes all natural substances – enable it to be functional matter – matter that serves the end of the substance it constitutes – and not merely a fully impressionable principle that can take on any form.[27]

Passages at the end of *Physics* I.9 contribute to the ambiguity of the analogy in *Physics* I.7:

> The matter comes to be and ceases to be in one sense, while in another it does not. As that which contains the privation, it ceases to be in its own nature; for what ceases to be – the privation – is contained within it. But as potentiality it does not cease to be in its own nature, for is necessarily outside the sphere of becoming and ceasing to be. For if it came to be, something must have existed as a primary substratum from which it should come and which should persist in it; but this is its own very nature, so that it will be before coming to be.[28]

In the following sentence, Aristotle defines material as 'the primary substratum of each thing, from which it comes to be, and which persists in the result, not accidentally'.[29] That it is a substratum from which natural things come to be which persists as material does not mean that it exists prior to the

---

[26] Jones's account of material as a certain relation seems susceptible to this 'characterless entity' view of material, 'Aristotle's Introduction of Matter', pp. 476, 488, 495, 500. Kosman argues that the relational account only works for artifice (p. 364), but still in conceiving of material as a principle collapses its character into being for form, as when he writes, 'At no stage do we find the distinction of matter into two beings, one with a nature other than that exhibited by that of which it is the matter, such as occurs in the case of artefacts, for at every stage the being of organs and their constituent elements alike are *naturally* devoted to their telic instrumentality', in 'Animals and Other Beings in Aristotle', p. 389.

[27] Gill, *Aristotle on Substance*, pp. 129-33.

[28] *Phys.* 192a25-31.

[29] *Phys.* 192a32-3. Zeller takes this passage to refer to prime matter. Robinson argues that if matter ceased to be it would arrive at itself, so he must be talking about prime matter or the elements, and since if the elements ceased to be and arrived at matter, there must be prime matter, he must presuppose prime matter in either case, 'Prime Matter in Aristotle', p. 175. William Charlton argues that this passage is not about lower materials, but rather making the point that it is not material, but the composite that comes to be and passes away, in 'Prime Matter: A Rejoinder', p. 198.

generation as the substratum. The man comes from seed, which is specific material which is transformed in generation. This specific material is required for it to be what it is. But this specific material also becomes functional material, material that is what it is for this substance to be – flesh and bones and tissues and organs for the human being. It nonetheless retains its relation to the elements, which is why it is always susceptible to decay. Gill argues that the generic nature of material as simple bodies is what remains potential in natural substances so that in this way it never ceases to be its own nature, and it is the simple bodies that while formed from one another are not made out of anything simpler.[30]

Those who defend a notion of prime matter would have to take the 'of each thing' (ἑκάστῳ) to be distributive as if each thing has the same material substratum instead of recognising that different things play the role of substratum in different substances. This case is one of many where they take mention of the way a specific substance acts as material in a process to mean some fundamental being is matter as such. But if the nature of such a thing is to be pure potentiality, containing a privation would *not* make it cease to be its own nature. The material ceases to be its own nature, which was the privation of the substance it comes to be. Against the advocates of prime matter who argue from this passage that prime matter is as much at work in the material – the seed from which the plant generates – as it is in the fully actualised substance, Jones argues that this view of matter as prime matter would make what underlies in natural things two.[31]

Whether this passage shows that material in generation is always already formed in some way and so privation indicates the sense of lacking the form that it will become or if it shows that material progresses from simple to more complex bodies – two readings which are consistent with one another – the material always has an initial nature. And it is this nature that ceases to be in generation, while in a certain sense persisting. Water ceases to be water when it becomes blood, and blood ceases to be blood when it becomes flesh; earth ceases to be earth when it becomes seed, and seed ceases to be seed when it becomes the plant. But water has its own nature, as do blood and flesh, earth and seed, even though they are also material for something higher up. Water still has a nature, understood in terms of the elemental forces of cold and wet or in terms of its directionality as in between. Aristotle continues that it still has its nature potentially when it has been further formed.

---

[30] Gill, *Aristotle on Substance*, pp. 109-10, 133-5.
[31] Jones, 'Aristotle's Introduction of Matter', p. 482 contra Code, 'The Persistence of Aristotelian Matter', p. 358.

The generation of natural substances from a ὑποκειμένων is more difficult to understand than the generation of artefacts because what the artefact comes from is more recognisable in the artefact that is formed, but these fundamental material elements also remain in natural substance, both wholly transformed and maintaining their elemental potentiality. A further difficulty is that substances themselves become the underlying thing. This positioning of the substance as the substratum is the case Aristotle makes in *Metaphysics* Z.3.

Metaphysics VII.3: Substance as Substratum

In a 1956 article, Hugh R. King contends that Aristotle's argument in *Metaphysics* Z.3 is that if matter is what remains when everything is stripped away, and substance is this substratum, then the indeterminate potentiality would be substance. Aristotle's point is not to discuss the nature of matter, King argues, but rather to argue that substance defined as that which is not predicated of a substratum is insufficient for explaining material.[32] Gill argues in the same vein that the point of this chapter is not to establish that substance should not be understood as substratum, but that substratum should not be understood as that which merely underlies. The substratum that characterises substance must be separable and a this. Aristotle proceeds by showing the problems of a substratum that merely underlies by positing a notion of material as that which remains when all else is stripped away, a case that is not Aristotle's view of matter, but one that allows him to show the problems with this conception of substratum.[33] Aristotle writes:

> The statement itself is obscure, and further, on this view, matter becomes substance. For if this is not substance, it is beyond us to say what else is. When all else is taken away evidently nothing but matter remains. For of the other elements some are affections, products, and capacities of bodies, while length, breadth, and depth are quantities and not substances. For a quantity is not a substance; but the substance is rather that to which these belong primarily. But when length and breadth and depth are taken away we see nothing left except that which is bounded by these, whatever it be; so that to those who consider the question thus matter alone must seem to be substance. By matter I mean that which in itself is neither a particular thing nor of a certain quantity nor assigned to any other of the categories by which being is determined. For there is something of which

---

[32] King, 'Aristotle without Prima Materia', pp. 387–8.
[33] Gill, *Aristotle on Substance*, pp. 25–37. Pace H. M. Robinson, who argues that the substrate that is laid bare is prime matter, which cannot be substance in Robinson, 'Prime Matter in Aristotle', pp. 184–6.

each of these is predicated, so that its being is different from that of each of the predicates; for the predicates other than substance are predicated of substance, while substance is predicated of matter. Therefore the ultimate substratum is of itself neither a particular thing nor of a particular quantity nor otherwise positively characterized; nor yet negatively, for negations also will belong to it only by accident.[34]

The concept of matter under consideration strips it of all potentialities suggesting that all potentialities belong to the material accidentally. This concept of matter is unlike the way that Aristotle describes matter in other contexts. It does not even seem like what one might call prime matter since it is without potentiality altogether – it seems like the receptacle view of material associated with Plato. As nothing, it would not be capable of doing the work of explaining generation from what remains. Thus, this concept of matter is one that Aristotle presents in order to reject. This passage has been linked to *Phys.* IV.2.209b5–15 and *Metaphysics* III.5.[35] In the *Physics* IV.2 passage, Aristotle presents a notion of matter as magnitude in space, a position which he associates with Plato's view in the *Timaeus* and proceeds to dismiss. Similarly, in the *Metaphysics* passage, Aristotle considers whether a body, a surface or a line is a substance in order to dismiss the argument raised by his predecessors. The parallel suggests that Aristotle is similarly raising this view in order to dismiss it. And indeed, Aristotle proceeds to dismiss this argument with reference to 'those who adopt this point of view'.[36] While this passage is taken to refer to those who think substance is a substratum, where that means what remains after all has been stripped away, that view of all else being stripped away depends on understanding material in that way, which these other passages dismiss. For material as that which all else has been stripped away would seem to be nothing. In *Metaphysics* H.1, Aristotle defines material as potentially a 'this', while the account in *Metaphysics* Z.3 makes it what remains when all else is stripped away.[37] This view of material as the boundless is the position of Anaximander's that Aristotle explicitly rejects to argue that instead it is the elements, which have some character, and are unbounded only in the sense that they can become the other elements.[38] At the end of this chapter, Aristotle does not focus on material, but rather on the role of separability and thisness, where form becomes the real complication, not material.

---

[34] *Meta.* 1029a9–26.
[35] Charlton, 'Prime Matter: A Rejoinder', p. 204.
[36] *Meta.* 1029a27.
[37] *Meta.* 1042a26–7.
[38] See *Phys.* 187a12–26.

Aristotle's argument is that substance cannot be a substratum if substratum is understood on a certain sense of material as what remains when everything else is stripped away. This version of a materialist account of nature can be attributed to those like Antiphon who think a bed is most wood, because if it could reproduce, it would reproduce a tree, not a bed, as Aristotle recounts his argument in *Physics* II.1. Antiphon ignores that the bed reproduces as a natural thing, not as artifice, and what is natural about it is what can generate on its own. Wood is a substance as a natural thing, which is a substratum. The example of the bed shows that wood is what a thing is when stripped down only on the model of artifice. But then on the account of artifice, if material not the composite is the substratum, then not even the bed is a thing in itself. Substance is a substratum rather as that which is separable and that which is a τόδε τι, a this. Aristotle can conclude this passage having transformed substratum into what is separable and a this recognising that form or the composite would be the substance. The composite seems derivative of form or matter, so it can be dismissed and form considered. Form however is difficult to think. Aristotle then claims that people agree that there are substances among sensible things. This is the difficulty – form might seem to be the substratum, and yet form existing in sensible things is difficult to think as the substratum.[39] This expression of the difficulty in terms of form testifies to Aristotle's concern with thinking substance as what is separable and hence a substratum for the other changes that occur within it, and hence as something more than just a bare characterless base.

De Caelo IV.5: The Elements in Common

One of the main passages that advocates of prime matter turn to is *De Caelo* IV.5. The Greek reads thus:

> Ὥστε ἀνάγκη καὶ τὰς ὕλας εἶναι τοσαύτας ὅσαπερ ταῦτα, τέτταρας, οὕτω δὲ τέτταρας ὡς μίαν μὲν ἁπάντων τὴν κοινήν, ἄλλως τε καὶ εἰ γίγνονται ἐξ ἀλλήλων, ἀλλὰ τὸ εἶναι ἕτερον.

J. L. Stocks translates this passage thus:

> The kinds of matter, then, must be as numerous as these bodies, i.e. four, but though they are four there must be a common matter of all – particularly if they pass into one another – which in each is in being different.[40]

---

[39] *Meta.* 1029a27–1029b12.
[40] J. L. Stocks, *The Complete Works of Aristotle*, *De Caelo* 312a30–2.

Guthrie translates the phrase that is taken to refer to prime matter, thus:

> in the sense, however, that there is really one matter common to all (because, for one thing, they are generated out of each other) but logically susceptible to differentiation.[41]

Dancy offers 'for in such a way that there is one common matter for them all'.[42] Solmsen, arguing that the four is a reference to matters, not elements, translates this passage as 'though they are four, there must be a common matter of all'.[43] Sheldon Cohen seems to follow that reading by translating the passage with plural matter:

> So the matters, too, must be as many of these [earth, air, fire, and water], that is, four, but four as one common to all (at least if they come to be out of one another), though the being of each is different.[44]

Commentators have argued that the elements themselves seem to need what Christopher Byrne calls 'a more basic physical stuff, whose *only* essential attributes are extension, mobility and corporeality'. Byrne is self-consciously trying to transform Aristotle's account of material into a Cartesian sense of an extended stuff.[45] His view, as well as those of Solmsen, Dancy and others, hypostasises a common matter rather than seeing the elements as what is common to all material substance. King argues that if it is the case that elements need some common matter that underlies to explain their change, then there is some moment when the prime matter has thrown off the previous form to take on the new form and thus in this in-between moment it would not be.[46] To avoid this possibility, Cohen argues that there is a common matter, but it is not bare, so it would not be nothing in the process of change between elements. Cohen argues that it is characterised by the contrarieties of hot, cold, dry and wet, which give positive significance to the common matter, but still makes prime matter reflect the Platonic receptacle: 'If we want it to be the common matter of the terrestrial elements it cannot be *per se* hot, cold, fluid, solid, light, or heavy, but must be capable of becoming any of these.'[47] Cohen's argument seems to be that

---

[41] W. K. C. Guthrie, *De Caelo*, p. 361.
[42] Russell Dancy, 'On Some of Aristotle's Second Thoughts about Substance: Matter', p. 389.
[43] Friedrich Solmsen, 'Aristotle and Prime Matter: A Reply to Hugh R. King', p. 249.
[44] Sheldon Cohen, 'Aristotle's Doctrine of the Material Substrate', p. 175.
[45] Christopher Byrne, 'Matter and Aristotle's Material Cause', p. 87.
[46] King, 'Aristotle without Prima Materia', p. 375.
[47] Cohen, 'Aristotle's Doctrine of the Material Substrate', p. 179.

the common matter is always characterised, but the contrarieties do not *per se* characterise common matter, otherwise it could not be potentially all the others.

But it is also possible that the elements themselves are the common matter of all things. To make that case, King suggests this translation:

> Thus it is necessary that there be the same number of [elemental] matters as these [classifications of matter according to gravitational pulls], four; and the four [elements] are as one, the common matter of all things, especially since they are generated from one another, but each is different in nature.[48]

King's argument emphasises the ways the elements are generated from one another to show them to be the common matter, rather than some common existent between them. William Charlton agrees with Solmsen's philological argument that the four indeed modifies material but the 'all', ἁπάντων, in this passage does not mean all materials then – since it is the four not the all that modifies material – but all things. He translates the passage: 'but four in such a way as to be the one common matter of all things'.[49]

This view that the elements are the common materials for all material substances is supported by passages earlier in *De Caelo* as well as elsewhere in Aristotle's corpus. At *Physics* IV.5, Aristotle says that water is the matter of air and air the actuality of water because water is potentially air and air is potentially water.[50] In *De Caelo* III Aristotle argues for a view of the elements as the basic existing things with natural motion that are generated out of one another. If there are no things with their own natural motion then the basic being of the world, not having order, would be disordered.[51] If the most basic components of the natural world do not have their own nature, then all movement is forced and no natural movement can be attributed to natural things whereby they move on their own accord. If only one thing has natural motion, then the multiplicity of motion cannot be explained.[52] Observation shows us that there are multiple motions. For these reasons, elements must have their own natural motion and they must be multiple. Charlton notes that Aristotle is here rejecting the notion of elements found in Plato's *Timaeus*, where elements are composed of space and differentiated by the addition or substraction of shapes.[53]

---

[48] King, 'Aristotle without Prima Materia', p. 384.
[49] Charlton, 'Prime Matter: A Rejoinder', p. 199.
[50] *Phys.* 213a1–3.
[51] *De Caelo* 301a7–11.
[52] *De Caelo* 304b12–14, 312b20–3.
[53] Charlton, 'Prime Matter: A Rejoinder', p. 206; *De Caelo* 306b16–21, 308b10–28.

Aristotle argues that the elements are generated and generate from one another on the basis of eliminating the other possibilities in *De Caelo* III.6. Aristotle defends the view that the elements are generated because they can be divided. Either they are infinitely divisible, or the division comes to an end. If the division ends, then the elements can be destroyed. Aristotle argues that the division must end, because if it did not, then the synthesis and the analysis of elements would proceed infinitely, which would mean there are two infinites, which Aristotle rejects. So there must be a stopping point, and the stopping point is not an atom, nor 'a divisible body which analysis will never reach' – which would seem to rule out something like prime matter – nor from something incorporeal, because that requires a void.[54] So Aristotle offers two possibilities: either elements are generated from something incorporeal, which is not possible, or from a body, which is also not possible, because then that body would be prior to the elements, but if it has any bodily existence (if it possesses weight or lightness), it would be one of the elements. If the elements have been established to explain the multiplicity of movement, this thing from which elements come into existence would seem to have no movement of its own but to explain what has movement. But if it has no movement of its own, it also would not have a place, so it seems that it therefore would not be a bodily existing thing, having no place. Aristotle rejects both possibilities – that it comes from something incorporeal, or from something corporeal and more basic than elements. He concludes, 'The elements therefore cannot be generated from something incorporeal nor from a body which is not an element, and the only remaining possibility is that they are generated from one another.'[55]

Having shown that the elements must generate out of one another, Aristotle proceeds to argue for real difference between the elements in order to affirm that they are multiple natures. That argument supports the view that change between them is generation and not merely alteration, which itself supports Aristotle's larger argument that generation in general is true generation and not merely alteration. In *De Caelo* IV.3, Aristotle considers the ways that elements are like one another and unlike in terms of natural movement: 'Now, that which produces upward and downward movement is that which produces weight and lightness, and that which is moved is that which is potentially heavy and light, and the movement of each body to its

---

[54] *De Caelo* 305a5. Pace Robinson who argues that the substratum ruled out here is a separable one, but Charlton notes that Aristotle here argues that if there is such an incorporeal substratum it will be separable. Such a substratum would be incapable of causing the generation and destruction of the elements. Robinson, 'Prime Matter in Aristotle', p. 177; Charlton, 'Prime Matter: A Rejoinder', p. 205.

[55] *De Caelo* 305a31-2.

own place is motion toward its own form.'⁵⁶ He explains that place is the boundary of the element which becomes in a sense its form, which is why one element seems to be the form of another because its place is the boundary of the other.

On this account of fire as the form of the whole cosmos based on its place, it would seem that fire is most akin to form as such because it goes toward the outermost boundary and earth to material as such because it goes toward the innermost boundary down to the centre. Aristotle argues that there are indeed things which are absolutely light as fire and absolutely heavy as earth to affirm that the elements exist in their own right.⁵⁷ In conclusion to this passage, Aristotle returns to his argument that form is the container and matter is what is contained, but here notes that the material can serve as material in one way for the heavy and another way for light:

> The same holds, consequently, also of the matter itself of that which is heavy and light: as potentially possessing the one character, it is matter for the heavy, and as potentially possessing the other, for the light. It is the same matter, but its being is different, as that which is receptive of disease is the same as that which is receptive of health, though in being different from it, and therefore diseasedness is different from healthiness.⁵⁸

The heavy and the light have a matter, which is that which is potentially heavy or light, but not yet. Earth is potentially fire, and so the matter of fire, and fire is potentially earth and so the matter of earth. Air and water both have some weight and some lightness and so can be the form and matter for one another. In conclusion, fire is no more form than earth, since both can potentially be the material of the other in the process of generation, here described in terms of place, but not wholly reducible to it, as we will see in *On Generation and Corruption*. Further, as fire goes up and earth down and air and water can move in between, the kinds of material must be as many as these kinds of movements. Here we return to the passage that opens this section making it clear that Aristotle is in no way setting up an argument for some fundamental underlying prime matter that makes generation and destruction possible.

---

[56] *De Caelo* 310a31–5; see also 311a19–311b29. Helen Lang argues from this passage that the form of the elements is the place toward which they tend in *The Order of Nature in Aristotle's Physics: Place and the Elements*.
[57] *De Caelo* 310a35–310b15.
[58] *De Caelo* 312a16–21.

How Matter is Many

In *De Caelo*, Aristotle argues that there must be multiple elements in order to explain multiple movement. In *On Generation and Corruption*, Aristotle argues that there must be multiple elements if generation is to be understood distinctly from alteration. On this basis, Aristotle dismisses the view that all things are created out of a single element, which is a fairly straightforward rejection of prime matter.[59] Rejecting prime matter is not only about arguing that material has its own character, but about distinguishing generation from alteration – 'the being of this matter and the being of alteration must stand and fall together'[60] – which allows Aristotle to think form and generation as more than arrangement and re-arrangement. The difference within material is what allows generation to be true transformative change and not merely alteration of some characterless entity. The multiplicity of matter also suggests that its capacities are positive rather than privative. For example, the argument that cold, which is typically associated with the feminine and viewed as a privation, has a capacity would seem to be a rejection of a view of prime material where some material is closer to prime matter because it is stripped of positive potentiality.

While those who think matter is only one end up making all change alteration, Aristotle notes that those who 'make the matter of things (ὕλην) more than one must distinguish coming-to-be from alteration'.[61] In *De Caelo*, Aristotle argues that if there is only one single material, whether that be the void or plenum or extension or triangles, then all material will act the same way.[62] If generation is more than just changing what was into something else by addition or subtraction, then what exists would have a distinctive quality of its own that distinguishes it from other matter. Yet if matter is more than one, and generation and alteration therefore different, then alteration seems impossible because there is no similarity between the elements allowing them to become one another. Alteration requires material to be fundamentally one and generation to be derivative of alteration. True generation requires material be multiple, but then generation would seem to create something out of nothing, which seems impossible.[63]

---

[59] *GC* 314a8–12, 320b22–3. C. J. F. Williams argues that this is inconsistent with positing prime matter and concludes that it demonstrates a confusion on Aristotle's account in Williams, *Aristotle's De Generatione et Corruptione*.
[60] *GC* 314b28–9.
[61] *GC* 314a11–12.
[62] *De Caelo* 312b21–3.
[63] *GC* 315b22–4.

The problem of matter being one or many is the problem of whether matter is a mere stuff and form creates the multiplicity of things or matter has discrete capacities that show it to have its own power. Another way to say this is that the problem of material being one or many is the problem of whether material is indivisible or divisible. If material is indivisible, no difference is material because it is not characterisable, and it is one. If it is divisible and having distinct character, it is many.

Aristotle raises this issue of material's divisibility in order to discuss action and passion in generation. There is a dispute among early philosophers regarding whether what acts – the agent – and what is acted upon – the patient – are identical or different. Aristotle argues that for change to be possible, the agent and patient must be distinct, and thus, matter needs to be divisible and multiple. Early philosophers accepted the Monists' view, Aristotle explains, that if there is to be motion, there must be a void, where void is not-being.[64] Positing the void is the way that Monists can maintain that material is one and change is real, but by positing the void, the Monists posit multiplicity: there is something other than what is. Insofar as there is something other than what is, what is is divisible. For Atomists like Leucippus the divisibility of the multiplicity arrives at an ultimate indivisibility of the primary constituents of matter.[65] Empedocles accepts the multiplicity of what is, which allows him to have a view of generation where there are agents and patients. Aristotle argues that Empedocles's view would make the elements themselves divisible because even the elements come-to-be and pass away.[66] Aristotle himself holds this position that material is multiple and that the elements are divisible, and the difference between the elemental forces allows generation of elements from the previous elements where the active and the passive roles of the elemental forces shift depending on what is generating and what is passing away.

For generation to be possible, multiplicity is required because the agent and patient must be different. For the agent and patient to be different, what is must be divisible. But if the material is a divisible magnitude, it would seem that it could be divided at every point simultaneously and thus be nothing or, at its most basic level, invisible.[67] Aristotle rejects this position of Democritus who thinks that the primary things are invisible indivisible magnitudes, because if no point is continuous to another point the magnitude is 'divisible through and through',[68] which means it would seem to be divided into nothing. It is not

[64] GC 325a26-9.
[65] GC 325b18-19.
[66] GC 325b21-2.
[67] GC 316b29-35.
[68] GC 317a3.

possible that the most basic component is either an indivisible or a divisible magnitude. If it is indivisible, no multiplicity is possible; hence no agent and patient for generation is possible. If it is indivisible that also means that it cannot be affected – it cannot be hot or cold or heavy or light or hard or soft or large or small.[69] All this to say, if matter is indivisible, it would have to be one, alike in every way with itself. Material would not be able to have any characteristics if it is indivisible. But if it is divisible, it would seem at base to be nothing.

Aristotle explains how material generation can be distinct from alteration and matter be multiple without being some specific divisible or indivisible building block through the language of actuality and potentially. The concepts of actuality and potentiality allow Aristotle to have multiplicity and generation. This move from generation as arranging material, where the elemental forces would be like properties appended to other forces, to actualising potentialities makes generation more than alteration, material multiple and potentialities of material multiple.

How we think of material as a part of generation is based on how we think of generation. Material is a building block if generation is association; material can have its own character if generation is the coming into being of a contrary from something potentially being its contrary from something that actually was something else. This sense of multiplicity stands in contrast to the building block notion of material that Aristotle finds in his predecessors who think of all generation as alteration, as the association of material things. It is a critique of the view that form is just the arrangement of a basic material stuff. Instead, generation and destruction are relations between what comes to be and that out of which it comes to be. As Aristotle writes:

> On the contrary, this is where the whole error lies. For unqualified coming-to-be and passing-away are not effected by association and dissociation. They take place when a thing changes, from *this* to *that*, as a whole. But they suppose that all such change is alteration; whereas in fact there is a difference. For in that which underlies the change there is a factor corresponding to the definition and there is a material factor. When, then, the change is in these factors, there will be coming-to-be or passing-away; but when it is in the thing's affections and accidental, there will be alteration.[70]

Unqualified coming-to-be and passing-away require that there be a difference between patient and agent, which makes what is in one sense divisible and in another sense indivisible. Change that is alteration is of accidents and points

---

[69] *GC* 326a1–10, 24–9.
[70] *GC* 317a19–26.

to the way that a thing is many and divisible. Change that is generation is of substance and points to the way that a thing is one and indivisible. The sense in which it is divisible and indivisible seems most evident at the level of elemental forces where the change in the level of what is active and what is passive manifests itself. That is, change at the level of the elemental forces explains how change at the level of the most basic material – elemental powers – could be generation and not alteration, without being association and dissociation.

The view of generation as the process of associating a material building block with other building blocks falls short because the building block would have to be either divisible or indivisible and both cases seem impossible. The alternative then is to understand material on the same model offered elsewhere as a relation, while at the same time recognising that it is its material capacities that allow it to be both agent and patient, an account which Aristotle articulates in *On Generation and Corruption* I.3 when he discusses the relation of the certain elements:

> Now we often divide terms into those which signify a 'this somewhat' and those which do not. And the issue we are investigating results from this; for it makes a difference *into what* the changing thing changes. Perhaps, e.g., the passage into Fire is coming-to-be *unqualified*, but passing-away-of something (e.g. of Earth); whilst the coming-to-be of Earth is qualified (not *unqualified*) coming-to-be, though *unqualified* passing-away (e.g. of Fire). This would be the case on the theory set forth by Parmenides; for he says that the things into which change takes place are two, and he asserts that these two, viz. *what is* and *what is not*, are Fire and Earth. Whether we postulate these, or other things of a similar kind, makes no difference. For we are trying to discover not what undergoes these changes, but what is their characteristic manner.[71]

Aristotle maintains that while we divide the material world into what looks more like substance and what does not, the goal of the investigation is what kind of change generation is. The kind of change will tell us more about what the material is than what the material is can tell us about what kind of change is at work.

For a material, whose constitutive differences signify more a 'this somewhat' (τόδε τι), is itself more a substance while a material, whose constitutive differences signify privation, is more not-being. (Suppose, e.g., that the hot is

---

[71] *GC* 318a32-318b9.

a positive predication, i.e. a form, whereas cold is a privation, and that Earth and Fire differ from one another by these constitutive differences.)[72]

Aristotle is not concluding that hot is a positive predication in contrast to cold which is a negative or privative predication, but, referencing Parmenides, saying that it is not that the one elemental force is more a building block, but that the one elemental force is more a τόδε τι, the language Aristotle uses for things that are a substance, 'a this'. Three chapters later, Aristotle approaches his argument for elemental forces as what underlies elements, when he explains that 'heat and cold do not change reciprocally into one another, but what changes (it is clear) is the substratum [τὸ ὑποκείμενον]'.[73] As Aristotle distinguishes the work of heat and cold in *Meteorology* IV:

> All this makes it clear that bodies are formed by heat and cold and that these agents operate by thickening and solidifying. It is because these qualities fashion bodies that we find heat in all of them, and in some cold in so far as heat is absent ... Of all the bodies that admit of solidification and hardening, some are brought into this state by heat, others by cold. Heat does this by drying up their moisture, cold by driving out their heat.[74]

This capacity of fire to thicken or solidify depends on whether it works on water or earth.[75] Throughout the whole of *Meteorology* IV, Aristotle continues to explain the work of hot and cold in light of the different ratios of the other elements that the hot and cold are working on. Aristotle changes the paradigm from one of building blocks to one of the actuality and potentiality, activity and passivity, where one elemental force is the actualisation and activity of one element and the potentiality and passivity of another element. The elemental forces change in relation to one another, where different forces remain in the change and as a composite define the element. When water as wet and cold transforms into air as dry and cold, cold remains, but in the transition the dry is what determines the difference from water, but the cold can determine the difference from fire, which the air has not yet become. In elements, what underlies as what remains can also be what most characterises the element. In the change then, there is a ὑποκείμενον that remains and the elemental force

---

[72] *GC* 318b14-18.
[73] *GC* 322b16-17.
[74] *Meteor.* 384b24-6, 385a22-3; see also 383a13-17.
[75] *Meteor.* 383b19-20.

that changes defines its change from what it was. The elemental forces shift their roles in the generation of elements. The elements themselves can be both a τόδε τι and a ὑποκείμενον, just like substance is, but the elements are transformed when they serve as a ὑποκείμενον in natural substance. Maintaining their character as elements when constituting natural substance is what makes natural substance have distinctive material character.

On Generation and Corruption

Despite Aristotle's outright rejection of prime matter as indivisible or divisible magnitude, supporters of prime matter argue that the transition between elements in elemental change is evidence that there must be some common matter that underlies and makes this change possible. The *De Caelo* passage suggests that the elements themselves are the common matter in all natural generation including of the elements. Prime matter supporters argue that a passage in *On Generation and Corruption* establishes a common prime material that underlies all elemental change, ultimately arguing that matter is one and not multiple. In the effort to distinguish between alteration and generation, Aristotle writes in *On Generation and Corruption* I.4:

> Since, then, we must distinguish the substratum, and the property whose nature it is to be predicated of the substratum; and since change of each of these occurs; there is alteration when the substratum is perceptible and persists, but changes in its own properties, the properties in question being either contraries or intermediates ... But when nothing perceptible persists in its identity as a substratum, and the thing changes as a whole (when e.g. the seed as a whole is converted into blood, or water into air, or air as a whole into water), such an occurrence is a coming-to-be of one substance and a passing-away of the other – especially if the change proceeds from an imperceptible something to something perceptible (either to touch or to all the senses), as when water comes-to-be out of, or passes-away into, air; for air is pretty well imperceptible. If, however, in such cases, any property (being one of a pair of contraries) persists, in the thing that has come-to-be, the same as it was in the thing which has passed-away – if, e.g., when water comes to be out of air, both are transparent or cold – the second thing, into which the first changes, must not be a property of this. Otherwise the change will be alteration.[76]

---

[76] *GC* 319b8–11, 14–24.

Further establishing what makes a change uniquely generation, Aristotle writes in the concluding passage of this chapter, 'but when nothing persists of which the resultant is a property (or an accident in any sense of the term), it is coming-to-be, and the converse change is passing-away'.[77]

Supporters of prime matter argue that this passage indicates that a perceptible part survives in nonsubstantial change which is alteration and an imperceptible part survives in substantial changes, taking 'nothing perceptible' to gesture toward the persistence of something that is imperceptible. They then take 'nothing' in the final passage of the chapter to mean 'nothing perceptible', and argue that something does remain but it cannot be perceived.[78] What remains that is imperceptible, they argue, is prime matter. Charlton argues that perceptible here does not mean 'capable of being perceived by the senses', but rather a perceptible object in contrast to an abstract object such as a mathematical object or a theoretical one.[79] It seems that elemental change poses a particular problem to which prime matter appears to be an answer because it would be what remains throughout the change, making it possible. The question is how elemental change is possible in a way that is not merely alteration. For the change to be generation of one element from another, 'the second thing, into which the first changes', that is, water from air, cannot be merely a property of the previous thing.

While supporters of prime matter argue that if water cannot be predicated on air, then there must be something else that underlies the change, this passage would be better understood to show that the transformation of elements does not occur from the addition of properties but rather from the shifting roles of the elemental forces in the transformation. In the generation of elements, one elemental force is not an added property, but they take on different roles as form or substratum to generate the new element. As Aristotle begins the next book of *On Generation and Corruption*, 'Our own doctrine is that although there is a matter of the perceptible bodies (a matter out of which the so-called elements come to be), it has no separate existence, but is always bound up with a contrariety.'[80] Against those who argue that there is a matter of perceptible bodies that is prime matter, Aristotle explains all change between elements in terms of the relational structure of the elemental forces that are the contrarieties. Aristotle has already established the notion that

---

[77] *GC* 320a1–2.
[78] See Robinson, 'Prime Matter in Aristotle', pp. 170–3 on how these passages support prime matter. Against this view, Gill, *Aristotle on Substance*, pp. 41–82.
[79] Charlton, 'Prime Matter: A Rejoinder', p. 207.
[80] *GC* 329a25–7.

elements serve as material for one another in *De Caelo*. In *On Generation and Corruption*, he continues to make the case that the structure of change between elements follows the structure of change in all other natural substances:

> But the elements must be reciprocally active and susceptible, since they combine and are transformed into one another. On the other hand, hot and cold, and dry and moist are terms, of which the first pair implies *power to act* and the second pair *susceptibility*. Hot is that which associates things of the same kind (for dissociating, which people attribute to Fire as its function, is associating things of the same class, since its effect is to eliminate what is foreign), while cold is that which brings together, i.e., associates, homogeneous and heterogeneous things alike.[81]

As the previous section establishes, Aristotle argues against the Pre-Socratics that the elements are neither building blocks nor made of more basic building blocks, but are constituted from the elemental forces, which are here described as the reciprocal capacities to act and be acted on, to be active and to be susceptible to activity. Hot is active by joining what is like and separating it from unlike and cold is active by joining the like and unlike. Aristotle addresses what he means by active and passive in generation in *On Generation and Corruption* I.7 in the context of whether material is one or many, which amounts to the question of whether change is between what is like or what is unlike, and more, whether any change can be true generation or whether it is all alteration. Two absolutely alike things do not affect one another.[82] Two absolutely unlike things cannot affect one another.

> But since only those things which either involve a contrariety or are contraries – and not any things selected at random – are such as to suffer action and to act, agent and patient must be like (i.e. identical) in kind and yet unlike (i.e. contrary) in species. (For by nature body is affected by body, flavour by flavour, colour by colour, and so in general what belongs to any kind by a member of the same kind, and it is contraries which reciprocally act and suffer action.) Hence agent and patient must be in one sense identical, but in another sense other than (i.e. unlike) one another. And since patient and agent are generally identical (i.e. like) but specifically unlike, while it is contraries that exhibit this character: it is clear that contraries and their

---

[81] *GC* 329b23–30; see also *De Caelo* 307a31–307b4.
[82] *GC* 323b3–9, 18–24.

intermediates are such as to suffer action and to act reciprocally – for indeed it is these that constitute the entire sphere of passing-away and coming-to-be. We can now understand why fire heats and the cold thing cools, and in general why the active thing assimilates itself to the patient. For agent and patient are contrary to one another, and coming-to-be is a process into the contrary: hence the patient *must* change into the agent, since it is only thus that coming-to-be will be a process into the contrary.[83]

If the patient changes into the agent, it must be like it in some way, just as the contrary of not-form changes into form, even though it looks like material changes into form in *Physics* I.7. Hot and cold and the moist and dry are alike in different ways. Hot and cold are alike as active principles of a certain kind. The cold dry thing and the hot dry thing are both dry. They are the same in species – both dry. But they are different in kind around temperature. When both are dry, the contraries around which the change occurs are not in virtue of what is the same, but in virtue of what is different, the cold to hot, though the dry and the moist can also be understood as opposed to the hot and cold. The active principle in the change is that into which it changes, that which is actualised, the hot. The cold dry thing becomes the hot dry thing. But here it is not the cold that is the patient that changes into the agent, it is the dry thing that is the patient.

Democritus argued that agent and patient were identical because a contrary being opposed to its contrary could not act on it. This problem is the one Aristotle addresses in *Physics* I.7, and he explains it here too:

> For sometimes we speak of the substratum as suffering action (e.g. of the man as being healed, being warmed and chilled, and similarly in all the other cases), but at other times we say what is cold is being warmed, what is sick is being healed: and in both these ways of speaking we express the truth, since in one sense it is the matter, while in another sense it is the contrary, which suffers action.[84]

Different elemental forces become susceptible to change in order to produce different elements. In *De Caelo*, Aristotle describes the character of elements in terms of place, where form is the outer boundary of the place to which they aim and the matter is that which is contained within that boundary. In *On*

---

[83] GC 323b29–324a14.
[84] GC 324a16–19.

Generation and Corruption, Aristotle fills out the characterisation of elements in terms of the elemental forces. The elements are composites of the elemental forces that have the capacity to act or be affected in different ways.

> Air, e.g., will result from Fire if a single quality changes; for Fire, as we saw, is hot and dry while Air is hot and moist, so that there will be Air if the dry be overcome by the moist. Again, Water will result from Air if the hot be overcome by the cold; for Air, as we saw, is hot and moist while Water is cold and moist, so that, if the hot changes, there will be Water. So, too, in the same manner, Earth will result from Water and Fire from Earth, since both tally with both. For Water is moist and cold while Earth is cold and dry – so that, if the moist be overcome, there will be Earth; and again, since Fire is dry and hot while Earth is cold and dry, Fire will result from Earth if the cold pass-away.[85]

Though the elements are composites of elemental forces, they get characterised with one elemental force chiefly. If the substratum is what most defines, then this cycle would make dry the substratum of earth, cold of water, moist of air and hot of fire, while the other elemental force would be the attribute of the element. To judge then whether a thing is essentially hot or cold or dry or moist is to judge which elemental force underlies, which we can judge by that into which it transforms.

In the cycle of change in *On Generation and Corruption*, some elemental forces are contraries, others are the substrata in a shifting cycle in each elemental change. When hot overcomes cold, earth becomes fire, when cold overcomes hot, air becomes water. In the shift from fire to air, hot is the substratum, and dry and moist are the contraries. In the shift from air to water, moisture is the substratum, and hot and cold are the contraries. Just as Aristotle says in a way fire is most form and earth most matter but that each element can be the form or matter for the others, similarly, moisture seems most like the substratum and heat most like the driver of change that is coming-to-be but each elemental force can be the substratum or the contrary in each elemental change. The elemental contraries continue to characterise the elements positively even when construed as passive. The character of the element varies depending on whether hot or cold works on moist or dry and how specifically that moist and dry is mixed to be able to be affected in a particular way to produce a particular material element. When the moist thing changes from hot to cold, the hot and cold

---

[85] *GC* 331a26–331b2.

are the contraries. But the hot or cold can also be what underlies as when the hot thing changes from dry to moist. In this way, the process of generation of the elements allows us to think of the same element as having different contraries in different changes, a situation that is unique to the elements and elemental changes since both components can play both the role of the contraries and the role of what underlies. The contraries can be understood as akin to form, but it would be wrong to think that the contraries are what give character to the element, since it is the relation of the contrary to what underlies that produces the specific character of the element, and this character can be produced regardless of which elemental forces act as underlying thing and which act as contraries. Elemental change is paradigmatic insofar as it shows how material contributes to the character of the natural substance, and how material itself has character that is not dependent on form.

This reading allows us to better understand how material plays a specific role in Aristotle's account of generation even as it has its own specific character. As Aristotle writes:

> Matter, in the most proper sense of the term, is to be identified with the substratum which is receptive of coming-to-be and passing-away; but the substratum of the remaining kinds of change is also, in a certain sense, matter, because all these substrata are receptive of contrarieties of some kind.[86]

Matter as the substratum can both come into existence and cease to be. That is an impossible thing to say of prime matter, but it could be said of the elements. If matter is meant only in terms of a relation here, if that is the most proper sense of the term, then this role is what can come into being and pass away. We have seen that to be the case, as when the wood ceases to be the material for the box and returns to being the wood beam that it was before being made into a box. But that does not quite seem to be Aristotle's point since he proceeds to discuss matter as a role in change in contrast to matter 'in the most proper sense of the term' in what follows. This passage then seems to describe a basic material that generates and passes away as part of its character, which would describe the elements.[87]

On this account, the elements are hylomorphic compounds where the elemental forces alternate in the role as form and material. The elemental forces

---

[86] GC 320a2–4.
[87] See Charlton, 'Prime Matter: A Rejoinder', p. 200; Gill, *Aristotle on Substance*. Pace Williams, *Aristotle's De Generatione et Corruptione*, p. 213.

themselves are the most basic material constituents but the roles they play are as that which underlies the change and that which characterises the generated element. This process requires no more fundamental material substratum.[88] This argument explains why some elemental changes occur more quickly than others, which prime matter cannot explain.[89]

Elemental change invites us to think of how the underlying part can be what remains even though it is transformed. Against the replacement view of the prime matter supporters, where one form replaces another in a stuff lacking any character of its own, Aristotle's account of change at the level of elemental change involves an underlying element that provides continuity as that from which the new substance emerges, but is also transformed in the generation of the new substance. Charlton notes Aristotle uses ὑποκείμενον when referring both to the sense in which accidents come to be in a substance as contraries from one another – musical from unmusical – and to things coming to be from what is not opposed (man becoming musical), but only ὕλη when the change is from what is not opposed. Charlton concludes that Aristotle does not suppose that what underlies must also remain as it was before the change.[90]

Returning to the initial passage that partisans of prime matter reference in *On Generation and Corruption* in light of this analysis shows how elemental change is at the centre of the dispute over the difference between generation and alteration. The hot is the substratum between fire and air, but the element as a whole changes when there is a shift from the contrary of dry to moist. Alteration involves a change where the substratum remains recognisable as itself distinct from the contraries that change. Generation by contrast is a transformation of the whole. While hot underlies in the elemental transformation, the element has become wholly new. The seed is transformed into blood in animal generation as water is transformed into air, Aristotle explains, because neither the seed nor the air is perceptible in what continues to be though a substratum persists. Hot and moist characterise air, but the moisture that characterises water is transformed when it becomes air.

---

[88] Eric Lewis argues that the elemental forces are like stuffs, the material components of the elements, and he equates this stuff with material and thus with a capacity to be substrata, a capacity that he argues is not in addition to being material, but just what it means to be material, which looks a lot like a common matter. Eric Lewis, 'Introduction', pp. 57–8.

[89] Gill, *Aristotle on Substance*, p. 74.

[90] Charlton, 'Prime Matter: A Rejoinder', p. 208; pace Robinson, 'Prime Matter in Aristotle', p. 177. Aristotle only suggests that what underlies must also remain in his attack on the view that there are things out of which numbers come into being. Gill distinguishes accidental from substantial generation on the basis of whether the ὑποκείμενον remains in *Aristotle on Substance*, p. 60.

Air shares moisture with water, but when water is formed from air, water is not a property of this moisture, by contrast to alteration as when musical is a property of man. The problem with those who support prime matter is that they understand generation in terms of alteration as merely changing characteristics of something that remains the same. Material as a substratum is understood as that which is without characteristics and form in generation is viewed as adding characteristics. Charlton argues that Aristotle would think it is 'mere gibberish' to speak of 'the form of fire' because fire and earth are expressions of matter for Aristotle. The reason we think we need to speak of form of fire is that we remain steeped within a tradition of reading form as what characterises a thing, and matter as what lacks character. But if material has its own distinct nature as material then we can speak of material distinctively without reference to form.[91]

Matter in Natural Substances: Generation as a Problem

These readings of *Physics* I, *Metaphysics* VII, *De Caelo* IV and *On Generation and Corruption* establish that material on its own terms – the elements – has a character of its own. Prime matter does not exist for Aristotle: there is no pure potential. Material, having its own character, poses certain problems for the unity of natural substance. One question is how material has its own character and becomes fully unified as the manifestation of the form of natural substance. This problem is resolved by recognising the difference between material on its own terms, which is elemental, and the material that is at work in the natural substance, both as the elemental and as what is transformed to embody the form. In the previous chapter, the discussion of form required some consideration of its relation to material in natural generation and substance. Similarly, this chapter's discussion of material requires some attention to form.

The other side to material having a character of its own is that natural form always exists within natural substance. Form for Aristotle is not hovering Platonic Forms that are instantiated in receptive matter. There is no universal man, Aristotle argues, 'Peleus is the cause of Achilles, and your father of you . . .'[92] Natural form is defined as that principle of movement within a natural substance.[93] Individual forms are both the source and the continued organising principle of natural substances. Natural generation appears to pose

---

[91] Charlton argues that Aristotle speaks of both form and matter as property-owners in 'Prime Matter: A Rejoinder', p. 209. Witt argues that form should not be understood as a property of substance, but a property of matter, *Substance and Essence in Aristotle*, p. 125.
[92] *Meta.* 1071a18-24.
[93] *Phys.* 192b14-21, 193b6-7; *Meta.* 1049b5-10.

a problem because in generation, the form and the material are separate and joined to form the new substance. In natural generation, the movement that most characterises natural things seems to look more like the motion of artifice than of nature because it is motivated from outside the substance. Natural generation looks like the patient needs to be activated by an external agent.

Yet while the principles of form and material are separate in natural generation coming as they do from the male and the female, neither of them are separated from form or material: as natural, the semen is already enmattered and the menses is already worked up to a certain point that gives it its capacity to become life. Moreover, γονή and καταμήνια are uniquely similar as form and matter so that the working of semen on καταμήνια is distinct from the working of the carpenter or even the tool of the carpenter on the wood. In persisting substance, natural form organises the elemental capacities to make the material fulfil the being of the substance and not move toward elemental aims. In natural generation, καταμήνια is similar to semen in its capacities, which is both rooted in material elemental force and capable of becoming animated because it has been worked up by vital heat. Semen has capacities through the material in and through which it operates. Καταμήνια is not in opposition to semen because they are similarly characterised except that semen is more concocted to be able to actualise life. Semen is not external and other as the form of a statue is in the mind of the artist. The καταμήνια is more like the patient who is potentially healthy and has the potential for health but needs to have health actualised. Καταμήνια does not resist γονή, it is already on the way to what semen actualises. The material in natural generation has its own principle toward generation, a principle which enables semen to do its work even as it is somehow made purposeful by the work of the form it carries through its motion. The γονή/καταμήνια distinction is a model of form and matter in natural substances because καταμήνια before generation is matter formed for the natural substance from a natural substance having its own principle of ordering. The matter for generation is like food for the nutritive soul, which is made food by being for the nutritive soul.

As I argue above, natural form or soul in natural substances is indistinguishable from the matter or body for Aristotle. In *De Anima*, Aristotle argues that the soul is substance in the sense of form of a natural body; it is the actuality of a body.[94] A living body is a body that is organised by a soul, so it makes no sense to question whether the soul and the body are one: 'it is as though we were to ask whether the wax and its shape are one, or generally

---

[94] *De An.* 412a20-1.

the matter of a thing and that of which it is the matter'.⁹⁵ Aristotle explains that the soul is the substance of a thing insofar as it is what it is to be the body that it is. This sense of being is explained in terms of the activity or function of a thing, the soul of an eye would be sight, the soul of an axe, chopping. This definition in terms of function also explains why the soul is not separable from the body.

Soul as the Actuality of the Body: The User and the Implement

At the conclusion of *De Anima* II.1, Aristotle says that despite this account of soul as the actuality of the body – as what it is to be this body – we do not know whether the soul is the actuality of the body in the sense that the sailor is the actuality of the ship.⁹⁶ On that model, the soul or form would still be separate from the body working separately from it to organise it. This model would reflect the model of artifice if that model can be understood as the form moving from the mind of the artisan to be imposed on appropriate material to become the artefact. But Aristotle argues that the soul is the movement, the purpose and the essence of the whole living body in a sense that removes the possibility that the soul is the actuality as a sailor on a ship is the actuality of the ship.⁹⁷ The analogy to the sailor might hold if the soul were the moving cause and the essence, since one could say that the sailor moves the ship and as the pilot of the ship contains what it means for the ship to be a ship. But the sailor is not the goal of a ship, sailing on water is the goal of the ship. The soul is the final cause of the body as what it means for the body to be this body. What it means to be a body is not in something else, but in the body, which is why Aristotle says that it makes little sense to ask whether the body and soul are one. For Aristotle, the soul is the power of self-organising and continuing to flourish that makes a living body what it is.⁹⁸

Because the soul is the actuality of the body, it does not make sense to speak of the soul without a body. The soul is not a body; it is 'relative to a body', σώματος δέ τι. Aristotle explicitly rejects the view that the soul could be 'merely fit into a body without adding a definite specification of the kind or character of that body' as if the body were the receptacle of the soul rather than the potentiality for the soul. Aristotle argues that the soul requires matter that is appropriate to it because 'the actuality of any given thing can only

---

⁹⁵ *De An.* 412b6–9.
⁹⁶ *De An.* 413a8–9.
⁹⁷ *De An.* 415b9–11.
⁹⁸ *De An.* 413b1–3.

be realised in what is already potentially that thing, i.e. in a matter of its own appropriate to it'.[99] In this passage, Aristotle is speaking of the material that is potentially a substance and the body that the soul actualises. The body is not stuff properly arranged by soul, but the organised manifestation of the living soul. Appropriate material is needed for this work of the soul. The appropriate material is 'already potentially that thing', the seed that becomes the living animal. Functional matter – the body as the matter fulfilling its function, fully formed – is distinct from the proximate matter that is potentially living, but not yet. Such material is on the way to the living being. Proximate matter is a good way to think of καταμήνια which is on the way to being the living material. It is not purely elemental, but its character is further organised and made appropriate for some higher form, but it is still distinct from the form that it manifests in natural substance. This matter is formed by a living being – the mother – and shares an end with the semen. Its capacities remain related to the elemental capacities of material, but become body when actualised by semen. Now actualised as functional matter as the body of a living soul, the elemental matter persists as potential, which is why the material substance is always subject to decay, which is a return to the elemental. Because the body remains potentially the elemental when actualised in the natural substance, the soul must continue to do the actualising unifying work to make the body functional matter, the matter for the unified substance.[100]

Aristotle describes the body as the tool or organ of the soul, πάντα γὰρ τὰ φυσικὰ σώματα τῆς ψυχῆς ὄργανα, in the context of explaining how the soul is the cause of the living body as the efficient, final and formal cause.[101] As in other natural things, the final and formal cause coincide:[102] the purpose and the ordering principle of the substance is the soul and its actuality. Having established that the soul is what it is for the body to be, Aristotle argues that the body is not just potentially soul, but the potential for the soul. The tool language suggests a separation of body from soul, but another way to consider that language is to see that the organ is what does the work of the soul, as fire is the tool of soul. Aristotle elsewhere says that deliberation considers the πρὸς τὰ τέλη of happiness, that by which the end is achieved, but commentators have argued that this is not merely the means, but could also be understood as that which constitutes

---

[99] *De An.* 414a26-7.
[100] See Gill, *Aristotle on Substance*, pp. 129-33, 146-65; Kosman distinguishes between these kinds of potentiality as between motion and substantial being, but I follow Gill on finding a parallel structure here based on making what is similar become what it is rather than the change where the contrary destroys the opposite.
[101] *De An.* 415b9-21.
[102] *Phys.* 198a25-9.

happiness.¹⁰³ Similarly, the organised body is the organ of the soul insofar as it is that through which the soul is manifested as soul. Understood on these terms, absorbing food is the manifestation of the psychic power of living.¹⁰⁴ The body is the organ through which food is absorbed. As Aristotle explains, 'what is fed is the besouled body and just because it has soul in it', and it only acts as food because what it is acting on 'has soul in it as a "this" or substance (τόδε τι καὶ οὐσία)'.¹⁰⁵ The body is the organ of the soul because the body is that whereby the soul can nourish the living being, but this organ- or tool-being of the body does not make it distinct or other than the soul. This relation of the body and soul explains why earlier in this chapter Aristotle says that soul, not fire, is the source of life. Soul is the organising force that organises by using heat or fire which through its organising force limits fire to be the source of this natural substance, rather than the element of fire.¹⁰⁶ But soul is not some hovering immaterial entity, but the limit of fire that aims it toward life.

The activities of the soul are in the body. The soul is what it is for the body to be the body that it is. Soul as actuality of the body explains why Aristotle describes the rational soul as not blended with the body – not because the activity of reason is somehow floating outside the body, but because reason has no bodily qualities. When the soul is not thinking, it is 'not actually any real thing' and it only is when it is engaged in that activity.¹⁰⁷ The activity is the manifestation of the rational soul, just as the nutritive activity of the body is the manifestation of the nutritive soul. Aristotle says Plato is right to call the soul 'the place of forms', not because the soul is carrying around separately existing Ideas, but because the soul is actualised by thinking of what is, so that 'this description [of the 'place of forms'] holds only of the thinking soul, and even this is the forms only potentially, not actually'.¹⁰⁸

When Aristotle analogises between the hand that is the tool of tools and the soul which is the form of forms, he would seem to suggest that the soul like the hand is that by which that which is actualised becomes actualised.¹⁰⁹ The soul

---

¹⁰³ *EN* 1112b12. See Richard Sorabji, 'Aristotle on the Role of Intellect in Virtue', and David Wiggins, 'Deliberation and Practical Reason'.
¹⁰⁴ *De An.* 416a19–24.
¹⁰⁵ *De An.* 416b10–11. J. A. Smith translates τόδε τι as 'this-somewhat', which makes τόδε τι seem like an instance of species form. I have modified the translation to 'this' to indicate the sense of the passage where food acts as food only insofar as that which it works on is independently directing itself to grow and increase, as the rest of the passage shows in Smith, *On the Soul* in *The Complete Works of Aristotle*.
¹⁰⁶ *De An.* 416a10–18.
¹⁰⁷ *De An.* 429a23–4.
¹⁰⁸ *De An.* 429a27–8.
¹⁰⁹ *De An.* 432a1–2.

can be a tool when tool means that by means of which something is actualised. But there is no sense in which forms *use* the soul, standing prior to it. The body manifests the soul, the rational soul manifests ideas, and hand manifests the tools in their work. It would be strange to say that the soul is the tool of forms if that meant that the soul is a mechanism separate from the forms, but not if it meant that the soul is that whereby the forms are actualised in being thought. Similarly, the semen as the tool of soul is that whereby and through which the soul works, where the soul is wholly immanent to it and not something that must hover outside it not itself at work in the semen. Rather, soul is what it is to have life in the καταμήνια, which is achieved through the semen.

Soul or the natural living form is what it is for the being to be alive. The goal of a living thing is for it to thrive as a living thing, which explains how the formal and final cause coincide. But Aristotle says the efficient cause too coincides with the formal and the final causes. That is easy to see in the mature natural substance, but how the efficient, formal and final coincide in natural generation is not as obvious because soul seems to work separately from the actualising that internally characterises natural substance, as the agent that works on the patient. The example Aristotle uses to illustrate that the three coincide is 'For man generates man (ἄνθρωπος γὰρ ἄνθωπον γεννᾷ).'[110] This example suggests that the soul that begets coincides in some sense with the soul that is begotten. The semen that is the tool of the father's soul is also that whereby the new substance actualises itself.

This difficulty of how the tool of the father's soul becomes the soul of the new substance is central to the questions of how semen works, and how form works in relation to matter in natural generation. The actual soul of the father produces that which becomes the potential soul of the new substance. Aristotle has been speaking of the way that body is the potential of soul and soul the actuality of the body. The soul is not an abstract force working at a distance, but the individual form that makes this body organised to be alive. The individual form of the father produces semen that becomes the potential soul of the new substance. The potential soul is the source of actuality, but not yet actual as life on its own, but it is the source of the actuality in the new embryo. The father's soul is actual in a different way than the semen actualises the new substance.

The Ways of Being Acted Upon: Like and Unlike

The different ways that the father and the semen act and actualise and that the semen and the menses are potential and acted upon require a consideration

---

[110] *Phys.* 198a27.

of Aristotle's distinctions between ways of being potential. In *De Anima* II.5, Aristotle distinguishes between three ways of speaking of the potential and the actual. There are those beings who could know by their nature, beings who could be knowing right now, and those beings who are currently knowing, the former are potentially knowing in different ways, but they are both ways that include the capacity to know. The one who is potentially a knower is the kind of subject who has the capacity to know though that capacity is not yet actualised.[111] Aristotle argues that the potential knower becomes an actual knower by being acted upon in a particular way by some other being that has already actualised her capacity to know. Aristotle argues that 'to be acted upon' can mean different things – it can mean an extinction of a contrary by its contrary or 'the maintenance of what is potential by the agency of what is actual and already like what is acted upon, as actual to potential'.[112] This transition is not one of alteration as it is generally understood, which is why a wise man is not altered in using wisdom.[113]

The reference to the destruction of a contrary by its contrary is to the previous chapter, where Aristotle describes the way that food is acted upon as the action of a contrary on its contrary. By contrast, the three relations of potentiality and actuality that Aristotle describes in *De Anima* II.5 are not the action of a contrary on a contrary, but this other sense of being acted upon that involves already being like, not opposed to, the source of change. Aristotle points to the dispute among his predecessors regarding whether food feeds what is like it or unlike it. The argument is that what is like cannot be affected by like and that change is always to its opposite. Aristotle rejects both of these claims in his view of what it means to be acted upon in *De Anima* II.5. The example of food appears to involve the transition from contrary to contrary. Aristotle acknowledges that change is always to an opposite, but he proceeds by complicating what is meant by opposite with his account of the different ways things are said to be acted upon. Crucial to his account here is that 'food is acted upon by what is nourished, and not the other way around'.[114] The point is that the body is not acted upon by food, but food is acted on by the body, which the food nourishes. The body uses the food because it is able to make the food, which is potentially like it, become like it. Food is unlike the body before it is digested. Once digested, it is like the body.[115]

---

[111] *De An.* 417a23–8. This view, supported by Gill, is contra the view that this kind of potential is a deprivation, *Aristotle on Substance*, p. 183.
[112] *De An.* 417b2–5.
[113] *De An.* 417b8–9.
[114] *De An.* 416a34–416b1.
[115] *De An.* 416b3–8.

Contrast the unlikeness of food that is acted upon by what is other than food – the body – that then transforms food into what is unlike, which is the body, with the likeness of the potential knower or student to the teacher or actual knower. The potential knower is acted upon as that which is like, being of the class of beings – τὸ γένος – who know.[116] Similarly, the καταμήνια is of the class of things that are potentially alive. Aristotle explains the first transition from potentiality to actuality in terms of generation and the capacity of the animal for sense, 'In the case of what is to possess sense, the first transition is due to the action of the male parent and takes place before birth so that at birth the living thing is, in respect of sensation, at the stage which corresponds to the possession of knowledge.'[117] The work of the male parent through γονή on the καταμήνια actualises the capacity for the καταμήνια to be a living sensing being. The καταμήνια is potentially sentient and becomes like the semen, which has the active potentiality for life and nutrition and sentience that is actualised in the καταμήνια. Gill argues that the same relationship between active potential and passive potential is found both in what Charlotte Witt calls the process and end (PE) relation of potentiality and actuality and the capacity and exercise (CE) relation of potentiality and actuality.[118] The καταμήνια is related to the semen in the first way as the passive potential to the active potential. The semen is actual as active potential rather than as actualised potential in the form of the father. In the second way, the living thing actualises its capacity to be fully living as the living thing it is – growing, sensing, knowing. As Aristotle explains: 'a thing may be said to be potential in either of two senses, either in the sense in which we might say of a body that he may become a general or in the sense in which we might say the same of an adult'.[119] The first potentiality is a potentiality in the nature of what it is to be a human being that has not yet been developed. The second is an option an actualised human being might put to work having developed the capacity for strategising. The first change is from a not-yet-developed potentiality that characterises the subject and the second is from a developed potentiality that is not currently at work. The first potentiality is a relation of semen to menses, where the menses is that which could be living actualised by the actual potentiality for life in the semen. The second potentiality is of the form actualised in material, of soul making the body be what it is to be that body, at work in the new substance. Both these accounts of potentiality involve a development toward what a thing is like, a transformation from a capacity to its actualisation.

---

[116] *De An.* 417a23–7.
[117] *De An.* 417b16–18.
[118] Gill, *Aristotle on Substance*, pp. 172–94; Witt, *Substance and Essence in Aristotle*, pp. 130–40.
[119] *De An.* 417b31–4.

Aristotle gives two categories for how things can be acted upon. The first is 'the extinction of one of two contraries by the other'. That would appear to be the way that food as the raw product is acted upon and transformed by the body, which incorporates it into itself, being otherwise than it is. The second is 'the maintenance of what is potential by the agency of what is actual and already like what is acted upon, as actual to potential'.[120] This is the way that the καταμήνια, being potentially alive, is acted upon by the semen which is alive.

The father makes the new animal capable of sensing, because the καταμήνια is the kind of subject that is capable of becoming a living thing, almost living on its own. As in teaching, the agent who is the teacher produces the change in the student to make the student more like the teacher by teaching the student, an action that is completed when the student learns. This motion is considered to be due to a difference between the patient and agent, and in a sense that is true, but their likeness is what Aristotle is more interested in when he describes their relation in terms of potentiality and actuality. Both the agent and the patient are knowers, one by having the active capacity to know, the other by being the kind of being who can know. The potential to know is active in the agent and passive in the student. This potential becomes actualised in the student who learns, and having learned is able to be a subject who can actualise that capacity at any time. There is some capacity and activity on the side of the student by virtue of being a student. The relation of the agent and the patient depends on sharing this potential to be a knower.[121] That relation of the student and teacher is the relation of the καταμήνια to the γονή.

Aristotle describes the relation of potentiality and actuality to explain a way of being acted upon on what is like by what is like it. Natural generation involves this kind of change, one where the material is similar to form, being potentially alive. The example of the doctor reflects a sense of likeness in what is acted upon. The patient must be potentially healthy, having the kind of body inclined toward health, to move toward the form of health that directs the doctor. Like the καταμήνια, the patient takes over the work of being healthy without the doctor. Aristotle argues in *Physics* B.1 that the doctor's aim is outside of itself, not toward more doctoring, but to health. The aim of nature,

---

[120] *De An.* 417b2-5. Note here that the potentiality to be acted upon, what is sometimes called a passive potentiality, is not without character or power. The parallel between the use of πάσχειν and πάθη shows how a way of being acted upon can be understood as a disposition for responding in particular ways, as Aristotle maintains in the *Nicomachean Ethics* and *Rhetoric*. See Marjolein Oele on the different senses and common strand running through Aristotle's notion of πάθη in her analysis of Heidegger's reading of Aristotle in 'Heidegger's Reading of Aristotle's Concept of *Pathos*'.

[121] *De An.* 417b5-7. See Gill, *Aristotle on Substance*, pp. 221-7.

by contrast, is towards nature. In natural things, a source of life – an ordering principle – acts on what is potentially and on-the-way to being life and continues to be at work in the living being because its goal is to actualise itself. In natural things, the capacity of that which is acted on to be acted upon follows from its similarity to what acts upon it. It is capable of becoming like what acts upon it because it is already like it. Καταμήνια and semen are already moving toward being alive, in a way that food is not yet food until it is transformed into what is otherwise than it is.

In the first potentiality Aristotle describes, an active potentiality acts upon a passive potentiality to make it like the active potentiality. This potentiality describes how the semen is the actuality of the καταμήνια in generation, which is different from how the builder is the actuality of the wood. The soul is considered the actuality or the user of the potentiality that is the tool of being. The temptation is to think that all generation involves a distinctiveness between the user and the implement, but the living substance generates in a way that makes the user and the implement more similar than not, both as part of the class of things that live, just as the student and the teacher are of the class of things that know. The soul and the body, the user and the tool, are indistinguishable in the living being. What is acted upon is not contrary to what acts. Καταμήνια is different than wood waiting to be organised, because καταμήνια becomes what it is in the same way that semen does though καταμήνια is not as far along. This makes καταμήνια like the student who strives to know but needs to have that knowledge actualised by one who already knows. The student is not without highly specified capacities.

If the craft model requires that some intermediate instrument makes something wholly other than the form in the mind of the artist into that form, the model of natural generation is that something that resembles the source as living works on something else that resembles the tool that does the work to make it like the source. One way of explaining the difference between the work of form in generation and the work of form in persisting substance is to argue that the work of form in generation begins in the κυῆμα. Then the material is already ensouled, already inchoate functional matter, so that form works similarly in natural generation as in natural persisting substance. But this account only pushes back the initial craft structure to a more basic stage, it does not solve the problem of how natural form and natural material work together to form a new substance.

A continuity between natural generation and the persistence of natural substance is required in order for form to be consistently working as natural form. For there to be a continuity, form and matter need to be always already intertwined. This intertwining is the Möbius strip model that best makes sense of the relation of form to matter in natural generation and substance.

This intertwining shows that material has its own power independent of form and that form's power is dependent on material, in a reversal of the traditional account that conceives material as wholly for form and characterised by form and form as independent of material.

Consider the differences of natural generation from artificial generation. In natural generation, the material that the substance comes from is transformed in the process initially when it becomes the material for menses from its elemental components and then in generation it becomes the material that constitutes the natural substance. The material in natural substance comes from a natural substance of the same form that the form comes from and both the material and formal contributions have been worked up in the same way, though differing by degrees. Natural generation and natural persistence seem different because natural generation seems to be a process of an external form imposing itself on material, but what is acted upon here is not contrary to what acts, but like it, sharing the potential to become what actualises.

This acting of like on like is the second way Aristotle describes how potentiality is related to actuality when the potential to come to know has already been actualised to form a being who is knowing.[122] This second way, Witt's CE, is the way that living beings are actualising themselves once the fetation is formed and is living. The emphasis in the literature has been on how the second form of potentiality is the way that Aristotle describes the potentiality in substance, which seems to distinguish it from potentiality in generation, allowing that such a potentiality seems like a relation of difference, which makes all generation seem like artificial making.[123] The distinction that makes the difference is between the ways of being acted upon, where the ways of being potential, insofar as they are establishing a likeness, describe natural substance, first in generation and then in the continued work of the soul to actualise the natural being. Both accounts of potentiality, both what Witt calls PE and CE, require a shared potential. This shared potential between form and material reaches back to the elemental capacities that underwrite material's power to become alive as well as semen's power to animate. Form can introduce an ordering mechanism and a purpose and still be interwoven with a dependency on material to do its work.

---

[122] Gill argues that the distinction at stake here is not between a potentiality for a state and a potentiality for an activity, a state of knowing versus the activity of theorising, for example, because Aristotle does not distinguish here between a potential knower and a potential theoriser, as he could have. Rather, both subjects have a potential for the same goal, a state of knowledge, which shows that potentialities are for the same end, though they can be had in different ways. Gill, *Aristotle on Substance*, p. 178.

[123] E.g. Kosman, 'Animals and Other Beings in Aristotle', p. 366.

4

The Feminine and the Elemental in Greek Myth, Medicine and Early Philosophy

The question of the relation of form to matter and of male to female in Aristotle must be understood in light of the historical and mythological context in which Aristotle exists and thinks. Even though Aristotle does not cite the Hippocratics directly, the work of the Hippocratics captures the views of health and medicine and bodies and gender that are contemporaneous with Aristotle. Regardless of whether there is any authorial unity to the corpus,[1] these texts help situate Aristotle's thinking and show what he is responding to within the accepted opinions of his time. Like Aristotle, the Hippocratics did not think in a vacuum. Helen King argues that the Hippocratic corpus must be understood within the broader Greek culture, extending back to Greek myth.[2] Paola Manuli maintains that the Hippocratics' accounts of women's bodies were not necessarily about how they observed women's bodies working, but their anxieties about how women's bodies work.[3] We might say that the Greeks' accounts of the female gods similarly reflect something about the anxieties and expectations about women and women's bodies held by those who wrote and perpetuated those stories, which is to say, Greek men.

This chapter considers women in Greek mythology and medicine not to deduce claims about what it was like for women in Greece on the basis of what

---

[1] For more on the history and composition and authorship of the Hippocratic corpus see Elizabeth M. Craik, *The 'Hippocratic' Corpus: Content and Context*.
[2] Helen King, *Hippocrates' Woman: Reading the Female Body in Ancient Greece*, p. 23.
[3] Paola Manuli, 'Fisiologia e Patologia del Femminile Negli Scritti Ippocratici Dell'Antica Ginecologia Greca', and 'Donne Mascoline, Femmine Sterili, Vergini Perpetue. La Ginecologia Greca Tra Ippocrate e Sorano', quoted by King, *Hippocrates' Woman*, p. 22. Andrew Stewart argues in 'Rape?' that 'Mythological pursuits and abductions represent nothing more or less than the projection of Athenian male desire first upon the heroic world and then upon the divine one.' Mary R. Lefkowitz is sceptical of this view, arguing that the Greek gods were not considered role models for mortals in *Women in Greek Myth*, p. 72.

the Greek goddesses were capable of or on the basis of how Greek doctors understood their bodily capacities. Rather, women in mythology and the treatment of women in medicine reflect the cultural assumptions and concerns of men about women. This chapter is not meant to be an exhaustive account of goddesses and women in myth and medicine, but to present a framework and a background for claims in fifth and fourth century BCE ancient medicine and natural science about gender differences by an examination of views of women's bodies and those elemental forces associated with them by drawing on select depictions of goddesses. In this first section of this chapter, I consider the role of female divinities in reproduction, with a consideration of those gods born out of Zeus and gods who reproduce without mating. I consider the ambiguity of the ἀρχή-ic role of Gaia as the source of living things, and the depiction of Pandora as the beginning of a separate race – women. Since it captures arguments criticised by Aristotle, I address the treatment of the role of woman in reproduction in the last play of Aeschylus's Oresteia cycle, *Eumenides*. Along the way, I show how the Greeks capture the fluidity and blood contribution of women. The second section of this chapter shows how Pre-Socratics thematise the elemental, which is then explicitly associated with gender in the Hippocratics as I show in the third section. The following chapters will reference this chapter to show the places where Aristotle is working out of general assumptions about generation and gynecological medicine and places where Aristotle is arguing against the generally accepted view.

Gender, Generation and the Gods

Greek theogony recognises that Greek goddesses are needed in reproduction. Whatever the playboy gods do and however independent they try to be, 'No Greek theorist ever argued that the male could provide this element [material] in reproduction.'[4] While Athena and Dionysus both appear to be born from Zeus, their birth requires their mothers' contribution before Zeus can intervene to take all the glory for himself. Athena is conceived when Zeus falls for Metis, daughter of Oceanus and Tethys, the personification of the clever. Metis keeps changing shape to escape Zeus, who finally penetrates her by changing shape to match her shape. When Metis is pregnant, her grandmother, Gaia, prophesies that Metis's first child will be a girl and the second will be a boy who will overthrow Zeus. Zeus's response is to find Metis and swallow her (after he tricks her into changing into a fly), just as his father had thought he could stave

---

[4] Lesley Dean-Jones, *Women's Bodies in Classical Greek Science*, pp. 151–2.

off the coming rebellion by swallowing his children. Note that Zeus does not swallow the offspring, but the mother carrying the offspring. Metis lives on in Zeus as his cleverness. The Greek men who write the myths imagine a pleasure in intercourse with a goddess (a pleasure both in sexual intercourse and being pregnant with), but recognise that such pleasure has a cost. Eventually, Zeus has a terrible headache and screams throughout the earth. The gods rush to help; Hermes realises the problem and has Hephaestus split open Zeus's head. Out of Zeus's head springs Athena, fully grown and armoured.

Zeus tries to manage the reproduction of Metis by swallowing her, but that power cannot be stopped, even though Metis dies. The emerging child requires an opening in the head of Zeus. By contrast to birth from a woman, Zeus cannot birth Athena on his own but requires the technological intervention of Hephaestus. The story of Athena's birth points to the fear of men that their offspring will overtake them, but also the recognition that this cannot be stopped. Zeus does not try to produce Athena on his own; he tries to prevent any offspring from coming forth in order to save his dominion, but the material forces of Metis cannot be stopped even by the king of the gods.

Dionysus is the other god who is born from Zeus. Like Athena, Dionysus is conceived in a woman. Semele, daughter of Cadmus, the king of Thebes, strikes Zeus's fancy, so he impregnates her, angering Hera. Hera makes Semele doubt whether Zeus is really divine so he promises to do whatever she asks. Semele asks him to reveal his divinity. Like an adolescent on a dare, Zeus cannot hold back. When he releases the thunderbolts and lightning and the earth shakes and fire is everywhere, it burns Semele. Zeus rescues Dionysus from Semele's burning body and puts the growing foetus in his thigh. When the baby is born, Hermes brings the baby to Semele's sister Ino and her husband Athamantas.[5] In this story, as in Athena's birth story, the child is conceived in the mother and only later moved into Zeus. The power of fire, which is by turns creative and destructive, destroys the mother.

Anxiety over the productive power of women is also found in Hesiod's original goddess of life, Gaia. While feminists might want to take Gaia as an emblem of power, she is not unlike Pandora as a locus of anxiety. The maternal earth that is its own source of life is a scary earth, likely to be just as reckless as those who might overreach her bounds. Without parentage and able to conceive by herself – Ouranos, the mountains and the sea are her children by asexual reproduction – Gaia is the double-threat of woman, unlimited by parent or mate.[6] If her power is in her capacity to give life and

---

[5] Apollodorus 3.4.3 and *Homeric Ode to Dionysus* II.1–21.
[6] Hesiod, *Theogony* 116–32.

this power is owed to no one, it is also why she is unrestrained from having her children castrate her partner, Ouranos, who lies on top of her with no chance to bring her Titan offspring into the world.[7] Ouranos, acting as a form that conceals without any place for the material possibilities to emerge, is confronted by Gaia, material force unleashed. Her source of life as her own makes her responsible to no one, except perhaps her children, and more to her daughter Rhea in her efforts to recover her own offspring than to her son Kronos, who keeps swallowing Gaia's grandchildren whole.

Hera jealously speaks of her capacity to give birth on her own in the *Homeric Hymn to Apollo*, and proceeds to bear Hephaestos and Typhon.[8] Hesiod tells us that Hera gives birth on her own to Hephaestos because she is angry at Zeus. In one version of the text, it is her strife that causes her to give birth. (This makes the story a bit strange because Hephaestos is the god who splits open Zeus's head to birth Athena, and yet it is this birth that angers Hera to go and conceive Hephaestos on her own. The gods are not bound by time.) Typhon was the giant who had a hundred serpent heads on each hand and was eventually imprisoned by Zeus in the deep darkness of Tartaros. Without men, the story suggests, women might give birth to monsters.

The self-conceiving woman was a serious threat, the bearer of head-splitters and monsters, the mother of castrators, plotters all. Like Gaia and Hera, Pandora is another mythical figure who captures the anxiety of the feminine among men. We learn first of Pandora from Hesiod, who describes her in *Theogony* as the forebear of 'wicked womenfolk' who are a 'nagging burden' to men, refusing to share in 'abject want, but only of wealth'. Hesiod accuses Zeus of having made 'women to be an evil for mortal men, helpmates in deeds of harshness', beginning with Pandora. Pandora brings the punishment of work without contentment. No matter what a man might do, the presence of woman is a burden. Women are the drones who do no work but eat their fill of what the worker bees produce. A separate race, women are unnatural, with an unnatural and wicked source.[9] Women make life harsh. But as Hesiod points out, there is no escape from the evils of women, because if a man escapes 'the malice of women', he will be alone in old age. The luckiest a man could be would be to marry 'a wife of sound and

---

[7] Lefkowitz calls the power to command loyalty a power peculiar to goddess mothers in *Women in Greek Myth*, p. 16.

[8] *Homeric Hymn to Apollo*, pp. 324–5; Hesiod, *Theogony* 929. See Lefkowitz, *Women in Greek Myth*, pp. 14–16.

[9] For more on woman as another race extending to Pandora, see Froma Zeitlin, *Playing the Other: Gender and Society in Classical Greek Literature* and Nicole Loraux, *The Children of Athena: Athenian Ideas about Citizenship and the Division between the Sexes*.

prudent mind', but even she would need her husband 'to balance the bad and good in her'.[10]

Nicole Loraux points to how Pandora in Hesiod's depiction is all appearance – more the things she wears rather than her body or substance.[11] King takes the appearance as deception further to argue that Pandora mirrors the deceptive oxen stomach that Prometheus presents to Zeus.[12] Like the stomach, which looked attractive but was just covering over bones, Pandora presents herself as παρθένος, maiden or virgin, when she is really γυνή, having already given birth, and now birthing uncontrollably the miseries of the world and women. King argues that following the model of the Hippocratics, who see the birth of the first baby as the end of the process of becoming woman, Pandora's deception as depicted by Hesiod in *Works and Days* is that she is γυνή all along and merely disguised as παρθένος. Pandora has birthed the grief, the woman, the ongoing mouths to feed, the endless production without limit that is the destruction of man. But she pretends to be the lovely maiden.

King finds this concern with deception regarding reproduction to pervade Greek men's view of women. If men think they plant the seed and woman is the field where the seed grows, men think as the sowers that they have the right to decide whether the seed should take root, but also that they lack control over the site in which the seed takes root.[13] King explains that male writers who suspect that women have ways of preventing conception are not indicating that women did in fact control and manage contraception, but 'this may be more appropriately seen as part of a wider fear in ancient culture that women have knowledge of drugs, herbs, and spells which are potentially damaging to men ... The myth of effective plant-based contraceptives may thus be a male expression of fear that women hold the knowledge which could enable them to control the fertility of the household.'[14] This effort to prevent women from controlling fertility might account for Hesiod's depictions of how Gaia, despite all her efforts, still does not control her fertility after castrating Ouranos: the bloody drops that fall upon her as a result of the castration impregnate her with the Furies, the Giants and the Melian nymphs.[15]

---

[10] Hesiod, *Theogony* 590–612.
[11] Loraux, *The Children of Athena*, pp. 80–1.
[12] King, *Hippocrates' Woman*, pp. 23–6. Lefkowitz argues that sexuality in general – which one could argue is ushered in by Pandora – was deemed dangerous because it deceives by affecting the mind and in particular, judgement, *Women in Greek Myth*, p. 23.
[13] This analogy is Page DuBois's focus in *Sowing the Body*.
[14] King, *Hippocrates' Woman*, p. 156.
[15] Lefkowitz, *Women in Greek Myth*, p. 17.

Men's lives are threatened by female reproduction because, just as Pandora is a beautiful image that conceals the destruction and suffering she ushers in, women's bodies conceal what cannot be controlled. Not only are women's reproductive organs internal and hidden, in the Greek organisation of public and private spaces women's bodies are hidden away from men. This concealment makes women potentially deceptive. Women threaten men with their reproduction, because, as Hesiod writes in *Theogony*, they both are the bellies and they produce the bellies they stuff from the fruits of the labour of others.[16] The word for belly, γαστήρ, can also mean womb, and as King notes, to have or hold in the γαστήρ means to be pregnant. Woman in Hesiod's poem is held responsible disparagingly for always being pregnant. Silvia Campese notes that woman both has a γαστήρ through which reproduction is made possible and she is a γαστήρ, a vessel, which must be controlled by a man because woman is so ravenous.[17] Pandora herself appears to be the exemplar of woman as the jar or womb, which is both voracious and productive, taking up the fruits of the earth and producing the demanding mouths.[18] As Hesiod writes in *Works and Days*,

> But the woman with her hands removed the great lid of the jar
> and scattered its contents, bringing grief and cares to men.[19]

The great lid of the jar of the womb has been removed for the necessity and the demise of men (not just man). And yet, women's refusal to open the jar or her control over it could be just as worrisome to men.

Hesiod does not generally seem to think that the feminine is only a jar, given the fecund power of Gaia in the *Theogony*. One might argue that mortal women, unlike goddesses, were merely jars. Aeschylus might be the one exception to the notion that no Greek theorist thought that the male could supply material in generation, when he maintains that the mother is merely the place of reproduction in *Eumenides*, the third play in the Oresteia cycle. Orestes has killed his mother, Clytemnestra, out of revenge for killing his father, Agamemnon, whom Clytemnestra killed out of revenge for sacrificing Iphigenia in order to move the winds to sail the ships to Troy to fight the

---

[16] Hesiod, *Theogony* 599.
[17] Silvia Campese, 'Donna, casa, città nell'antropologia di Aristotele', p. 16.
[18] King, *Hippocrates' Woman*, pp. 25–6; Guilia Sissa, *Greek Virginity*, pp. 154–5; Zeitlin, *Playing the Other*, p. 65. See D. Panofsky and E. Panofsky on Klee's 1920 drawing on Pandora's jar of a goblet resembling Fallopian tubes in *Pandora's Box: The Changing Aspects of a Mythical Symbol*.
[19] Hesiod, *Works and Days* 91–6.

Trojan War as told by Euripides in *Iphigenia in Aulis*. The cycle of revenge must be broken before the city is awash in blood. It stops by not shedding a man's blood. Throughout the cycle, it is evident that the shedding of women's blood underwrites and enables the definitive projects of the community. In the sacrifice of Iphigenia, Agamemnon contributes a sacrifice to the city, a sacrifice born of the menstrual blood of the mother.

King points to the linkage between menstrual blood and sacrificial blood where both are public blood – blood that is required for the success of the community.[20] To say the woman is fulfilled in the child is to say that she has put her blood to work for its purpose which furthers the community. This connection between menstrual blood and sacrificial blood is also found in the connection between marriage and death in Greek myths. Iphigenia is brought to Aulis to be sacrificed by Agamemnon under the pretense that she is to be married to Achilles. Plutarch tells a story in which a group of παρθένοι strangle themselves, an event which is explained with reference to Hippocrates, who maintained that the lack of flow led to strangulation.[21] If defloration is postponed too long, the παρθένος would be strangled to death by the lack of flow of blood. A woman is saved by intercourse, which opens up the flow, while woman is the source of the flow of blood that saves the city in the case of Iphigenia. Iphigenia's sacrifice makes the service of the public effort to go to war supplant the family contribution to the city in the form of reproduction. King observes that girls in Greek myths use the girdle that is supposed to protect their virginity to strangle or hang themselves after being raped. While her husband would be expected to release the girdle in defloration, the woman takes the power to manage her girdle and thus her body herself by using the girdle as a noose. Just as sacrificial death substitutes for marriage for Iphigenia, Antigone also transforms the girdle of the παρθένος into a noose in the place of marriage. Artemis, the goddess of the παρθένοι, was herself called the strangled goddess.

Agamemnon is responsible to the city to do whatever is necessary to move the winds. Iphigenia is responsible to the city and to Agamemnon to give her blood for the city to go to war. Clytemnestra is responsible to her husband, to support him. But his failure to maintain his responsibility to Iphigenia as his blood, blood that is less binding according to the logic of Apollo in Aeschylus's play because she is his daughter and not his son, leads Clytemnestra to bleed him in his bath. Orestes then takes the responsibility that he has to the blood

---

[20] King, *Hippocrates' Woman*, pp. 81–9.
[21] Plutarch, *Moralia*, 'On the Virtues of Women', 249b-d.

of his father as a demand for the blood of his mother, even though his very being owes its existence to the blood of his mother, as the Chorus says (hasn't he already taken it once before?).[22]

The question of whether Orestes in *Eumenides* is responsible for a crime in killing his mother hinges upon whether he owes her anything. The Chorus, Furies awakened by the ghost of unavenged Clytemnestra, says Orestes must pay because he killed the one who gave him birth. When Apollo asks the Chorus whether Clytemnestra, too, had a responsibility to Agamemnon, the Chorus responds that it is not the same kind of responsibility that Orestes had to Clytemnestra, as blood of her blood. Apollo denies this claim, while the Chorus calls for further blood to avenge the mother's blood soaked into the ground. Orestes says the sacrifice of swine is recompense enough. Here the sacrifice of the animal replaces the blood of the mother. The Furies insist that bloodshed must be met with further bloodshed, and that their purpose is to assign proper lots for human action,[23] and describe their task as to 'drive from home those who have shed the blood of men'.[24] The Furies agree to give this task to Athena,[25] which seems to be the moment that hearkens the end of the revenge cycle set up by the Furies. The end of this revenge cycle falls finally on the sacrifice of Clytemnestra, who is never avenged, and becomes, as the shed blood for which no blood will be shed, a sacrifice that enables the peace of the community and the end of the bloodshed. Just as Iphigenia's shed blood sets the Greeks in action to their glory in Troy, so the shed blood of Clytemnestra leads to the possibility of Greek action at home, freed finally from the responsibility for avenging blood. The capacity to decide falls to Athena who considers herself without mother (though as I said above, we know she had one), which makes her more willing to allow the matricide to go unpunished. In this move, she is depicted as wise and just. The Chorus asks if Orestes does 'forswear your mother's intimate blood?'[26]

If the play is taken as a moment of an overcoming of destiny (in the Furies) for the sake of justice (in Athena), then it also depicts the subordination of the motherblood to the patriarchal line. Nancy Tuana marks a shift from the 'female principle of creation [as] a metaphor woven throughout Greek cosmogony' captured in the figure of Gaia to metaphors of sowing and planting as the product

---

[22] *Eumenides* 151-2.
[23] *Eumenides* 211-13, 261, 284, 321-96.
[24] *Eumenides* 421.
[25] *Eumenides* 434-5.
[26] *Eumenides* 608.

of men's labour on the property of women's bodies.[27] This language is found in sexual reproduction in Hesiod[28] and later in Aeschylus, who describes Oedipus as one 'who sowed/his outrageous agony/in the inviolate field/of his mother',[29] and Sophocles who in *Antigone* has Creon defend killing his son's betrothed because 'there are other fields for him to plough'.[30] C. D. C. Reeve maintains in his notes on Aristotle's *Politics* that Athenian fathers would give their daughter in marriage with the words, 'I give you this woman for the plowing of legitimate children.'[31] Tuana marks this as a shift from a fertile earth that produced enough without labour to an agrarian culture that required working the land. Where life springs forth without work, it is feminine. Where it requires work, it is masculine.[32]

While there is considerable dispute in the literature over whether there is in fact a pre-historical matriarchy, it is sufficient to show that there are such references. A pre-historic or pre-Olympian feminine deity would not necessarily point to a matriarchal world. Such a deity could just as much capture anxieties about nature and its unpredictable whims of fertility. Nonetheless the projection back to such a deity and the sense that the current order of masculine gods had set things right still indicates concern with the uncontrollable, unmanageable role of women in reproduction, in the possibility of male immortality held in women's hands. The recognition of such a primordial goddess could make the patriarchy feel more fragile and less natural, and thus even more anxious about enforcing its supremacy.

Further evidence of the insistence of privileging the masculine is the criticism of how Agamemnon dies. Apollo argues that dying (and thus killing) in war is more honourable than dying (and thus killing) in one's private home, where one should feel protected.[33] Women are sneaky, conniving. Even an Amazon would fight like a man. But this death was not manly. As Hesiod tells us, a man expects to find comfort in a woman, and here is betrayed. The Furies call attention to how Zeus himself challenged his father, and so there is little support in Greek myth for the view that the father must have pride of place.

---

[27] Nancy Tuana, 'Aristotle and the Politics of Reproduction', pp. 190–1.
[28] Hesiod, *Works and Days* 810–13.
[29] Aeschylus, *Seven against Thebes* 956–9.
[30] Sophocles I 569. See DuBois, *Sowing the Body*.
[31] C. D. C. Reeve, *Aristotle: Politics*, p. 383.
[32] Related to this point is the convergence of autochthony and masculinity as described by Loraux, *The Children of Athena*.
[33] *Eumenides* 625–8, 632–5.

In the famous passage referenced by Aristotle, Apollo responds by challenging the notion that a mother is a parent at all:

> The mother is no parent of that which is called
> her child, but only nurse of the new-planted seed
> that grows. The parent is he who mounts. A stranger she
> preserves a stranger's seed, if no god interfere.
> I will show you proof of what I have explained. There can
> be a father without a mother. There she stands,
> the living witness, daughter of Olympian Zeus,
> she who never fostered in the dark of the womb
> yet such a child as no goddess could bring to birth.[34]

Aeschylus has Apollo invoke Athena as evidence that men can conceive without women but not vice versa. And yet, as we have seen, Greek myth did not depict men reproducing without women, only the other way around. If anything, Zeus is the one who is merely the carrier for both Athena and Dionysus. Aeschylus was surely not unaware of this. Yet Athena decides on the basis of this argument from Apollo on behalf of Orestes. Athena's response is that 'There is no mother anywhere who gave me birth.'[35] That her mother did not give her birth is not quite the same thing as saying her mother contributed nothing to her, since Athena is conceived in Metis, and merely carried by Zeus. Athena is a false example of Apollo's claim – her mother was not the place of her.

While the cycle of blood-letting among the mortals ends by absolving Orestes, the danger that the Furies would still bring vengeance by preventing the fruitfulness associated with the mother remains. The Furies have given their role of deciding destiny in this case to Athena, who appeals to them with reference to Gaia not to bring a curse on fertility in response to the decision:

> Earth be kind
> to them, with double fold of fruit
> in time appointed for its yielding. Secret child
> of earth, her hidden wealth, bestow
> blessing and surprise of gods.[36]

---

[34] *Eumenides* 640–4, 658–66.
[35] *Eumenides* 736.
[36] *Eumenides* 944–8.

Athena stops the cycle of revenge in three ways: first by taking the decision away from the Furies (the cosmic consequences of matricide are suspended), second by deciding for Orestes (against his mother) and third by appealing to the Furies to continue to foster fertility (the work of the mother). Gaia and her hidden wealth that is the power of fertility is here invoked. The capacity of the feminine earth to withhold fertile crops is repeated in the Olympian Demeter. Her power and the anxiety that she might withhold it are captured in her response to the sexual (mis)conduct of Hades, who has taken the fruits of Demeter's own labour, Persephone, to the underworld. The anger and arbitrary whim of women is associated with the whim of the earth to bring forth anew, a whim that must be managed and constrained.[37]

Ouranos tries to prevent the fertile power of Gaia starting a cycle continued by Zeus, who swallows Metis in an effort to destroy the offspring in order to preserve his rule. Zeus's child with Metis, Athena, appeals to the Furies, who demand revenge for maternal blood spilled (even as Athena leaves unavenged the blood of her own mother). Athena establishes an end to the bloodshed by taking over the duty of the Furies, replacing revenge with justice decided by some wise judge. The Furies express their anger and their intention to make the land barren, and even though justice has replaced fate, it is still the threat of the thunderbolt from Zeus – Athena's father and only remaining parent – that brings the Furies in line. With the threat of Zeus in the background, Athena appeals to the Furies to put persuasion in its rightful place and to take their rightful place in the household and out of public life. In this place, the Furies will no longer cut men down before their prime,[38] but instead have the power to help παρθένοι set up house with men. Aeschylus thus depicts the restraint of the powerful female Furies within the limits of the home, an accomplishment borne out of the sacrifice of Clytemnestra at the hand of Orestes.[39]

Aeschylus might be capturing an emerging conflict between the view depicted by the later gods to suggest that the woman contributes only place and the earlier gods like Gaia who are considered the source of life. Athena's appeal to the earth would seem to mark a recognition that earth is the source of life and not just the place for it. Further support for the play as a conflict between these views is that it opens with the Pythia, the spokesperson for

---

[37] *Homeric Hymn to Demeter*. See Zeitlin, *Playing the Other*, pp. 164–7. For more on Greek tragedy exhibiting male anxiety about female sexuality, see Nancy Sorkin Rabinowitz, *Anxiety Veiled: Euripides and the Traffic in Women*.

[38] *Eumenides* 825–32, 885–900, 956–7.

[39] See Luce Irigaray, 'Body against Body: In Relation to the Mother'.

Apollo at Delphi at a place that is also associated with the earlier mother goddess Sybil. The Pythia's opening line is, 'I give first place of honor in my prayer to her who of the gods first prophesied, the Earth; the next to Themis, who succeeded to her mother's place of prophecy.'[40] That pride of place is something to be feared and managed, replaced with male principles of justice and limitation and proper order. Aeschylus does not ignore that the cost of the institution of these principles is borne by women from Iphigenia to Clytemnestra. Women's power is co-opted by becoming *for* the masculine principles.[41]

There is no disputing that women contribute to reproduction, but that contribution does not elevate women's power in society; it increases men's sense of their need to control women. Women in their reproductive roles are considered deceptive and capable of withdrawing their cooperation beyond the control of men. Moreover, the Oresteia cycle as well as stories of gods born out of Zeus show how women contribute fluid and material in the form of blood to public life and private life in the blood of Iphigenia and Clytemnestra and to reproduction – Zeus can be the place but not the material that produces Athena and Dionysus. In these stories, women are not only material, but particular ways of being material. Women from Sybil/Gaia to Demeter, Artemis and Pandora are associated with earth and moisture or fluid in contrast to the fiery power of men, exemplified in the thunderbolt and lightening of Zeus and the forming power of Hephaestus. The elemental role in generation captured by gods and goddesses, heroes and heroines becomes conceptualised by the Pre-Socratics. As Adorno writes, 'ancient concepts are essentially secularized gods'.[42]

The Elemental among the Pre-Socratics

The purpose of this chapter is to contextualise Aristotle's work in the milieu of ancient Greek thinking. The discussion of the feminine in mythology shows a tension between a recognition of the power of women to procreate and an anxiety about how that power could change and challenge the place of men within their world. It also points to the ways that earth, blood and material

---

[40] *Eumenides* 1–3.
[41] Lefkowitz argues that women are depicted in Greek tragedy in a way that recognises their importance in Greek society, and that women's passivity sequestered from the action depicts the human condition of powerlessness in relation to the gods, an argument that would still seem to align masculinity with powerfulness and the divine, in *Women in Greek Myth*, p. 51.
[42] Theodor W. Adorno, *Metaphysics: Concepts and Problems*, p. 85.

are attributed to the feminine as a source of life and reproduction. In this section, I move backward historically from some of the commitments of the Hippocratics regarding the elements and material forces like balance and flow to trace those ideas to the work of Pre-Socratic thinkers. By looking at the Pre-Socratics we can see how Aristotle's work is situated within several centuries of disputes about how to understand not just change in general but the specific change of generation.

The Hippocratic authors reject the Milesians' view that nature in general and human beings in particular are composed at bottom of a single element, of air or fire or water or earth.[43] They also reject the view that human beings are a particular humour, like blood, which changes when influenced by the hot and the cold. Both the argument regarding humours and the argument that human beings are reducible to one element depend, one Hippocratic author argues, on the view that the human is a unity, that it is made of one thing, which could not be the case because a unity could not suffer pain and there could not be a cure because the cure would have to be one, but there are in fact many cures.[44] Rather, 'For in the body are many constituents, which, by heating, by cooling, by drying or by wetting one another contrary to nature, engender diseases; so that both the forms of diseases are many and the healing of them is manifold.'[45] This view draws on the Anaxagorean and Empedoclean view that there are many constituents that are condensed and rarefied to form different entities. The author of *Nature of Man* continues that if the human being is blood then there would be a season of hot or cold when a person could be blood alone, as influenced by the temperature, but since this never occurs, human beings are not just one humour.[46]

The Hippocratic author of *Nature of Man* argues for the multiple constituents that comprise natural beings, including humans, and each component 'must return to its own nature when the body of the man dies, moist to moist, dry to dry, hot to hot and cold to cold'.[47] The multiple elemental forces of moist, dry, hot and cold are the most fundamental. The differences between them are the source of the differences in humours. Pain is the excess or defect of any of the multiple forces or an isolation of a humour in a particular part of the body, moving away from the place where it was well-balanced.[48] Thus,

---

[43] *Nature of Man* I.1–18.
[44] *Nature of Man* II.2–8, 12–14.
[45] *Nature of Man* II.17–21.
[46] *Nature of Man* II.21–9.
[47] *Nature of Man* III.19–23.
[48] *Nature of Man* IV.1–20.

as both Heraclitus and Pythagoras suggest, health is rooted in establishing a harmony – relaxing what is tense and making tense what is too relaxed.[49]

The Hippocratic author of *Regimen*, referencing the work of various Pre-Socratics, considers the belly the source of power in the body, which organises the elements in relation to the elemental forces:

> The belly is made the greatest, a steward for dry water and moist, to give to all and to take from all, the power of the sea, nurse of creatures suited to it, destroyer of those not suited. And around it a concretion of cold water and moist, a passage for cold breath and warm, a copy of the earth, which alters all things that fall in it. Consuming and increasing, it made a dispersion of fine water and of ethereal fire, the invisible and the visible, a secretion from the compacted substance, in which things are carried and come to light, each according to its allotted portion. And in this fire made for itself three groups of circuits, within and without each bounded by the others: those towards the hollows of the moist, the power of the moon; those towards the outer circumference, towards the solid enclosure, the power of the stars; the middle circuits, bounded both within and without. The hottest and strongest fire, which controls all things, ordering all things according to nature, imperceptible to sight or touch, wherein are soul, mind, thought, growth, motion, decrease, mutation, sleep, waking. This governs all things always, both here and there, and is never at rest.[50]

The belly (the jar associated with the earth and the feminine in mythology) is the source of the moisture and 'ethereal fire', which will return in Aristotle's account of the source of heat. As there, fire is related to moisture and air, 'the power of the stars', and a mixture between them (which is how Aristotle will define semen). While Thales's claim that all things are made of water is dismissed by the Hippocratics and the philosophers who follow, Aristotle traces this insight of Thales to the observation that both seeds and what nurtures is moist 'and that heat itself is generated by the moist and kept alive by it', so water seems to be the source of all things, even heat.[51] Aristotle reports the view that moisture nourishes fire as a view that is widely shared in *Meteorology* II.2.[52] This connection between Thales and heat, and the intermingling of the elements returns in *De Anima* and *Generation of Animals*, where Aristotle explains that

---

[49] *Nature of Man* IX.4–10.
[50] *Regimen* I.X.1–26.
[51] *Meta.* 983b20–4.
[52] *Meteor.* 354b33–355a7.

living things come into being in earth and water 'because there is water in earth, and air in water, and in all air is vital heat, so that in a sense all things are full of soul'.[53] This last phrase is a reference to Thales that Aristotle mentions explicitly at *De Anima* I.5, pointing to the view that some people think that soul is throughout the cosmos, and for this reason Thales thought the world was full of gods.[54]

Thales is caricatured as the thinker of water, but the few references to Thales we have show that his account connects some elemental capacity of water to other elemental forces, such as heat, and to soul. Anaximenes assigns the role Thales assigns to water to air, and as for Thales, Anaximenes's position is not naïve. As for the Hippocratics and Aristotle, Anaximenes maintains that the elemental forces of cold and hot and moisture do specific work, which is to cause movement.[55] Hot and cold are the most fundamental components of generation because they are the cause of rarefaction and condensation.[56] Despite these common themes, Aristotle reduces Anaximenes's contribution to 'Anaximenes and Diogenes make air, rather than water, the material principle above the other simple bodies.'[57]

Aristotle draws attention to the view of Anaximander that opposites are present in and separated out of the one, and to Empedocles and Anaxagoras who suppose that there is one and many because things are separated out of a mixture.[58] Simplicius explains Anaximander's view to be that because the four elements changed into one another, none of them could be a substratum, which would not seem to be a stable underlying thing, so that had to be something else, what Anaximander calls the ἀπείρων.[59] For Anaximander, the elemental forces such as the moist and dry, hot and cold are most fundamental, and the ἀπείρων looks like the prime matter that many scholars locate in Aristotle's metaphysics. Anaximander traces the source of the generation of the celestial bodies (at least the sun and moon) to hot and cold which form the sun and moon out of the ἀπείρων.[60]

Anaxagoras, following a century later, offers a similar view to Anaximander, though he denies the intermediate role of the elements and makes what is that is

---

[53] GA 762a19-20.
[54] *De An.* 411a8-9.
[55] Hippolytus, *Ref.* I.7, 1 (DK13A7) in *The Presocratic Philosophers*, pp. 144-5.
[56] Plutarch, *de Prim. Frig.* 7, 947f. (DK 13B1) in *The Presocratic Philosophers*, p. 148.
[57] *Meta.* 984a5, Kirk, Raven and Schofield translation, *The Presocratic Philosophers*, p. 14.
[58] *Phys.* 187a20-3.
[59] Simplicius, *Phys.* 24, 21, in *The Presocratic Philosophers*, pp. 128-9.
[60] Ps-Plutarch, *Strom.* 2 (DK12A10) in *The Presocratic Philosophers*, pp. 106-7.

mixed and dissolved the ἀπείρων, the infinite or indeterminate in which everything has a part of everything else within it.[61] More importantly, Anaxagoras introduces the organising role of νοῦς, or mind, that which is unmixed and transcends that which is mixed and organises it, setting up precursors to Aristotle's hylomorphism.[62] Recalling the role of Zeus in Hesiod and thinking (νοεῖν) in Parmenides, Anaxagoras associates reason with the cause and principle of what is.

Thales, Anaximenes and Anaximander are addressing a question later thematised by Parmenides regarding whether change is possible. Parmenides maintained that there is only what is, what is cannot be related to what is not, and therefore, there can be no coming into being, which would associate what had not been to what is now. For Parmenides, what is and what remains the same is what is thought, νοεῖν. The Milesians address this question by saying that what is most is the elemental. Change occurs, but what is remains as it is – the elemental.

Given the Hippocratic concern with the role of the elements and elemental forces, Empedocles seems like one of the most Hippocratic of Pre-Socratics, and closest in time, historically. Aristotle argues that Empedocles is the first to make the material 'elements' four.[63] Connecting the gods to the philosophers, Empedocles speaks of the elements in a way that makes them very much personified and active, calling the four roots of all things 'shining Zeus, life-bringing Hera, Aidoneus [Hades], and Nestis [Persephone] who with her tears waters mortal springs'.[64] Zeus is fire, Hera air, Hades earth, and Nestis water. The personalities associated with each make this account of material far from a passive substratum.

Like Anaximenes, who argues that condensation and rarefication are the causes of things coming to be and passing away, and Anaxagoras, who argues for a mixing and dissolving guided by Νοῦς, Empedocles recognises equal and opposite causes that generate and destroy by joining and separating, and he calls these Love (Φιλότης) and Strife (Νεῖκος).[65] Love is both destructive and productive, destroying the individual being of a thing by joining it to others to make something new. Strife too is both destructive and productive, destroying what had been joined and creating something new. As Empedocles writes, 'Double is the birth of mortal things and double their failing.'[66]

---

[61] Anaxagoras, Fr. 1 in *The Presocratic Philosophers*, pp. 357–8.
[62] Anaxagoras, Fr. 12 in *The Presocratic Philosophers*, pp. 362–4.
[63] *Meta.* 985a31–3.
[64] Fr. 6, Aetius I, 3, 20 in *The Presocratic Philosophers*, p. 286.
[65] Fr. 17, Simplicius, *Phys.* 158, 1, 13 in *The Presocratic Philosophers*, pp. 289–90.
[66] Fr. 17, Simplicius, *Phys.* 158, 1 in *The Presocratic Philosophers*, p. 287.

The continuous movement of Love toward the centre and Strife toward the peripheries leads to a constant cycle of generation.[67]

The association of Love and Strife in Empedocles reminds us of the strife that follows from coupling in Hesiod's works. There is a sense in which joining itself produces the conflict that then leads to the separation, and the separation itself cannot last – what is drives itself back toward all the other things that are, which leads to further coupling, which then leads to divisiveness again (unconscious uncoupling, one might say). Empedocles sees the fundamental bases of material existence in the four elements. Aristotle seems to follow the Hippocratics in seeing the elements themselves as composed of elemental forces.

Anne Carson points to the way that the association with wetness and dryness becomes valued.[68] Diogenes of Apollonia associates intelligence with dryness: 'Understanding is the work of pure and dry air. For moisture hinders intelligence, wherefore in sleep and in drunkenness and in surfeit understanding is diminished.'[69] (Intelligence will become associated with coolness in Aristotle.)[70] Carson argues that emotion, anxiety and fear were associated with wetness for the Greeks from Aeschylus to Sophocles to the Hippocratics.[71] Heraclitus calls a dry soul wisest and best (σοφωτάτη καὶ ἀρίστη).[72] Whether Heraclitus makes this judgement against water and wetness is not so obvious since another fragment, one which makes him sound much more like Empedocles, suggests that soul is just another of the elements: 'For souls it is death to become water, for water it is death to become earth; from earth water comes-to-be, and from water, soul.'[73] While the dry soul is wisest, Heraclitus's soul seems to come from water, in a way that will be reflected in Aristotle's account of generation, where soul is in the semen that is comprised of water and air, the source of the heat in semen.

Heraclitus seems to agree with Anaximenes and Empedocles (and eventually, Aristotle) that the elements change into one another. The source of the

---

[67] Fr. 35, Simplicius, *De Caelo* 529, 1 in *Phys.* 32, 13, 3–17 in *The Presocratic Philosophers*, pp. 296–7. Aristotle takes issue with Empedocles's account of the relation between Love and Strife in the forming of the cosmos because Empedocles could not explain the immobility of the Earth (*De Caelo* 295a29). Aristotle is concerned that Strife seems too much like chance in Empedocles's account, and so it can't explain how in fact certain elements move to the places they do (*GC* 344a1).
[68] Anne Carson, 'Putting Her in Her Place'.
[69] Carson's translation, 'Putting Her in Her Place', p. 138. See DK 64 A19 in *The Presocratic Philosophers*, pp. 447–8.
[70] *PA* 652a35–7.
[71] Carson, 'Putting Her in Her Place', p. 138.
[72] Heraclitus, Fr. 118 in *The Presocratic Philosophers*, p. 203.
[73] Heraclitus, Fr. 36 in *The Presocratic Philosophers*, p. 203.

process through which this happens on Heraclitus's terms is through fire. Fire catalyses the transition between opposites: Heraclitus calls god 'day night, winter summer, war peace, satiety hunger', saying that god changes as fire takes on the scent of spices when it is mixed with them.[74] Fire turns the sea into earth.[75] And fire is that for which all things can be exchanged, just as gold is exchanged for goods, making it, like money, that through which things pass and the measure of those things.[76] Heraclitus references fire (and Zeus, the divine ἀρχή or measure) when he says that the thunderbolt steers all things,[77] and that the sun, the 'brightest and hottest', maintains the proper boundaries.[78]

The Pre-Socratics pose the questions that lead Aristotle to articulate the need for a formal and material cause (Aristotle offers four causes, but explains that in natural things, the efficient, formal and final causes coincide). The essence and the stuff of a thing both explain what it is. The Pre-Socratics articulate how change between elements can be explained and how particular elements drive that change. If Aristotle points to the need for a formal and material cause to explain what is, the Pre-Socratics point to these dual roles in the elements of fire and moisture, which it turns out, will return in Aristotle.

Hippocratics, Elemental Powers and Gender

The Pre-Socratics set the background to the role of elements in the Hippocratics and the philosophers of the fifth and fourth centuries BCE. But how do the elements become gendered? This section draws together Pre-Socratic claims with Hippocratic conceptions of gender. The notion of wetness as feminine could be traced to the conception of Pandora as the leaky jar. Carson argues that the wetness of woman is connected to the view that woman lacks boundaries in contrast to the dry boundedness of men.[79] Women are the content, what is bounded, fluid and wet, while man is the form, what bounds. In Hesiod, the innate wetness makes women's sexual appetite insatiable: because women are wet, they don't need to stop. Because they lack σοφροσύνη, they do not think to stop. The wetness of women's bodies, in Carson's words, 'threatens the essence of a man's manliness'.[80] Carson explains Hesiod's description of woman in *Works and Days*, 'The voracious woman, by her unending sexual demands,

---

[74] Heraclitus, Fr. 67 in *The Presocratic Philosophers*, p. 190.
[75] Heraclitus, Fr. 31 in *The Presocratic Philosophers*, p. 197.
[76] Heraclitus, Fr. 90 in *The Presocratic Philosophers*, p. 197.
[77] Heraclitus, Fr. 64 in *The Presocratic Philosophers*, p. 197.
[78] Heraclitus, Fr. 94 in *The Presocratic Philosophers*, p. 201.
[79] Carson, 'Putting Her in Her Place', p. 139.
[80] Ibid. p. 142.

"roasts her man" in the unquenchable fire of her appetite, drains his manly strength and delivers him to the "raw old age" of premature impotence.'[81] Here, the woman is fire as well as moisture. As Carson puts it, woman is more associated with the elemental, and so with the δύναμις of nature. As a leaky jar like Pandora, woman is an unbounded sieve that needs proper boundary. Marriage binds a woman to a man, bounding the material overflow of woman.[82]

The view of woman as wetness needing binding might seem to contradict the Hippocratic notion that sexual intercourse permits a flow that would otherwise strangle a woman, since here marriage becomes that which binds up an excessive flow. But it is possible that the flowing and the binding occur in marriage and motherhood. As wet and fluid, women need to flow, but they cannot manage their own elemental nature; something external is required to prompt the flow and to keep it within its bounds.

King argues that the wetness of women pervades the Hippocratic corpus.[83] In *Airs, Water, Places*, women are described as wetter than men.[84] Females incline toward water; males toward fire, which is why women thrive from cold and moist food and drink and men from hot and dry food and drink. As the Hippocratic author writes in *Regimen* I, for a man to produce a girl offspring, he needs to have more water; for a boy, he needs to take up a regimen with more fire.[85] The Hippocratics recommend intercourse just after menstruation for those who want to conceive because then the uterus and connected passages are empty so the seed would not be overcome with moisture.[86] As Dean-Jones notes, there was only one day in each month when the womb was dry enough

---

[81] Ibid. p. 141.

[82] Ibid. p. 143. See Xenophon, *Oeconomicus* 3.7-10.

[83] King argues that commentators have tried to produce distinct authors for the different texts in the Hippocratic corpus in order to explain differences and variation either in terms of different times texts were written or different authors who wrote them. Another way the discrepancies were explained was in terms of 'schools' of medicine, with one at Cos and one at Cnidos. King argues that the notion that a text could have authority without clear authorship was an embarrassment to classical scholars, so they sought ways to indicate authorship. For all this, King argues, the perceived discrepancies in the text are not discrepancies but different recommendations for the virgin and the adult woman in *Hippocrates' Woman: Reading the Female Body in Ancient Greece*, pp. 65-6.

[84] *Airs, Water, Places*, 10.

[85] *Regimen* I.27.1-9.

[86] As the Hippocratic author writes in this passage, 'Now if the fire fall in a dry place, it is set in motion, if it also master the water emitted with it, and therefrom it grows, so that it is not quenched by the onrushing flood, but receives the advancing water and solidifies it on to what is there already. But if it fall into a moist place, immediately from the first it is quenched and dissolves into the lesser rank.' *Regimen* I.27.18-25.

to conceive a boy.[87] When Aristotle argues that it is not too much moisture, but too much matter that prevents conception, he is associating matter with moisture while rejecting the notion that it is the moisture that is the problem for conception; rather the materiality, which is moist, is.[88]

Breath, too, for the Hippocratics, as for Aristotle, plays a crucial role in generation. In *Nature of the Child*, the Hippocratic author describes the process of reproduction as the mingling of male and female seed, which becomes thicker when it is mixed. Because the process occurs in a warm place it is able to take in the mother's breath as breath and when it is filled with this breath, a passageway out is formed, which becomes the umbilical cord. Through the passage, cold breath is drawn from the mother and warm breath exits. The author writes that all things that are warmed take in breath, and the umbilical cord, which is formed through this path out, allows it to take in nourishment. The warmed seed grows from the blood that comes into the uterus from the mother – blood is drawn in along with breath through the umbilical cord. The congealing blood forms flesh.[89] In this way, the Hippocratics join the work of air and moisture and heat.

Too much moisture for the Hippocratics can mean too much heat, which can harm a woman. During pregnancy, blood flows each day to increase the flesh of the seed as it grows. A woman who is not yet pregnant can sense whether the blood within her is cold or hot because she is moister. When the blood is stirred up and fills her vessels, some separates off. If a woman is empty, she becomes pregnant, but if she is full, she does not. If the blood that was stirred up stays in the uterus and the uterus becomes warm and makes the rest of the body warm this can lead to lameness in the woman. The lack of flow can make the woman's body infirm.[90]

The Hippocratic author describes the process of growth after conception through the work of breath, which increases the flesh. Breath moves parts to other parts that are similar to it, 'the dense to the dense, the rarified to the rarified, and the moist to the moist'.[91] While the Hippocratics use opposites to heal by contrast to Asklepios, in development like associates with like to

---

[87] Dean-Jones, *Women's Bodies in Classical Greek Science*, p. 170. Regimen I, 27.1-27.
[88] GA 727b10-11, Dean-Jones, *Women's Bodies in Classical Greek Science*, p. 186. Ann Ellis Hanson argues that Aristotle's discussion of wetness and women's health was not original, but all over the Hippocratic Corpus in 'Conception, Gestation, and the Origin of Female Nature in the *Corpus Hippocraticum*'. On her account, Aristotle associates the female with moisture in a context in which this view is taken as the background assumption.
[89] *Nature of the Child* 1, 3.
[90] *Nature of the Child* 4.
[91] *Nature of the Child* 6.

grow.⁹² Heat solidifies the bones, making them hard. 'Each of these things is articulated by breath, for through the force of blowing all things separate according to their kind.'⁹³

The main problem the elemental forces present in the body in generation is the possibility of excess of any particular one. Moisture can feed heat. So does breath. The woman's body that is full of moisture can become a problem. While intercourse might open the pathways to flow to let moisture out as mentioned above, another solution was in a woman's own body. The Hippocratic author of *Gland* suggests that the breasts' task was to draw off excessive moisture from the woman's body. The loss of a breast would lead to death because the breast would be unable to draw off moisture.⁹⁴

In *Places in Man*, the Hippocratic author allows that different people have different constitutions – some hotter and some colder. Pain is produced by that which is opposite to each person's general constitution. In the cold, pain is produced by heat; in the dry by wetness.⁹⁵ As King writes in a way that resonates with Socrates's efforts to cure Charmides in Plato's dialogue of the same name with a charm that turns out to be a dialogue about self-knowledge, 'To know your painkiller, you must know yourself.'⁹⁶

Conclusion

Some dispute remains among scholars of the Hippocratics and Aristotle's biological works over how much Aristotle was familiar with their work and drawing on it. Bartoš articulates the Hippocratic contribution as the notion that fire and water are the elemental principles, where fire moves and water nourishes.⁹⁷ The

---

[92] King, *Hippocrates' Woman*, pp. 103–4. King contextualises the differences between Asklepian doctors and Hippocratics, noting the reasons patients might seek out religious medical advice instead of Hippocratic help. Some patients may have preferred Asklepios to Hippocrates because Hippocratic practitioners might ask questions about the past that could be embarrassing but Asklepians would not. But the Hippocratics may have been more effective. Temple treatment was perceived as more pleasant than the Hippocratic, p. 111. For these reasons, King objects to the opposition between the religious model and the Hippocratic model in the scholarship and finds them more often to go hand in hand. She argues against drawing the wrong conclusions from absence of evidence, p. 105.
[93] *Nature of the Child* 6.
[94] *Gland* 17.
[95] *Places in Man*, 42.
[96] King, *Hippocrates' Woman*, p. 121.
[97] For a review of the history of the debates, see Hynek Bartoš, 'Aristotle and the Hippocratic *De Victu* on Innate Heat and Kindled Soul'.

Hippocratics inherit the concept of vital heat from the Pre-Socratics, which then influences Aristotle. Dean-Jones argues that 'menstrual blood is the linchpin of both the Hippocratic and the Aristotelian theories on how women differed from men'. For both the Hippocratics and Aristotle, women 'could not convert to their own cause' the blood or moisture which characterised them, though breath seems to do significant work in women in generation for the Hippocratics. The difference between Aristotle and the Hippocratics is that for the Hippocratics, there was 'no natural analogue in a man's body' to blood, which was the source of all women's problems. Its excess 'led women to pass their lives teetering on the brink of ill health', and the monthly draining of it kept women from dying. But Aristotle takes the moisture of menses to correspond to the heat of semen, though he describes menses and semen as developing through the same process. In contrast to the Hippocratics, Dean-Jones argues, Aristotle does not think the production of menses benefits the women's body.[98]

Several themes seem to emerge from the study of Greek mythology, Pre-Socratics and Hippocratics that return in Aristotle. Women pose a problem to the ancient Greeks who organise their families and communities in terms of patriarchal power where men dominate public life and women are secreted in private. The problem of women arises because within these assumptions of male superiority, women are clearly crucial to the work of reproducing society, which seems to uniquely empower them. Women are associated with moisture from Hesiod through the Hippocratics. For Hesiod, this association makes women's sexual desire insatiable. The wetness of women's bodies threatens to overtake men's bodies. As wet, woman is like a leaky jar. Aristotle in associating women with moisture is picking up on a cultural understanding that also associated men with fire. The moisture of women could be generative and excessive, just as the fire of men could be generative and destructive. Both were capable of consuming what was before it.

The Pre-Socratics point to various elements or all the elements together to explain continuity and change in nature. The Hippocratics say natural things have a multiple material basis, which explains how they can suffer and be healed, a notion that Aristotle draws on, also traceable to Anaximander and other Pre-Socratics, in considering the structure of acting and being acted upon. Like the Hippocratics, Aristotle understands elemental forces as fundamental forces, though he takes them to always exist in elements. The Hippocratics carry over from mythology the notion of the belly as a source of power in the body, which Aristotle adapts by considering the stomach as the earth of the

---

[98] Dean-Jones, *Women's Bodies in Classical Greek Science*, pp. 225–6.

body. Aristotle similarly seems to take from the Pre-Socratics the notion that natural things are formed from an interaction between material elements and organising forces, as when Aristotle concludes that the world is full of soul. Pre-Socratics like Anaxagoras establish a distinction between what is indeterminate and mind, in a way that influences a tradition of commentary that sees material as the indeterminate and form as intelligible.

5
Semen, Menses, Blood: Material in Generation

Aristotle is thinking about generation in the context of medical practices that construe the body in terms of elemental principles, all of which could be found in the body, some more commonly associated with male bodies and some with female bodies. The Hippocratics understood blood as the female contribution to generation. This blood could be excessive or deficient. For the Greeks from Hesiod to the Hippocratics, the moisture of the female is to be monitored and managed for the sake of generation: too much moisture can be harmful, but moisture can also feed heat, which is associated with the male body. For the Greeks, the stomach is the source of a strong fire that controls and regulates the body, and heat in general solidifies. The Hippocratic author of *Nature of the Child* explains the nutriment of the child *in utero* to come from breath. In the adult, breath moves parts of the body. The notion that a monstrosity results from the female overpowering the male can be traced to Hesiod's depiction of the offspring Gaia births from the blood that falls on the ground after castrating Ouranos. In this context, Aristotle lives, thinks and develops an account of how the male and female contribute to generation that incorporates these general notions of male and female bodies, even as he transforms those views to fit into his four-fold causal framework.

Aristotle remarks in *GA* II.1 that the seed (σπέρμα) has the principle of motion that brings life into parts animating them so that they become the parts that they are. For Aristotle both male and female contributions are considered seed for the sake of life. In *GA* I.18, Aristotle describes how sperm (γονή) comes to have that principle, beginning as it does as nutriment, just as the feminine contribution does.[1]

> Now a boy is like a woman in form, and the woman is as it were an impotent male (ἄρρεν ἄγονον), for it is through a certain incapacity that the female is

[1] *GA* 725a4–6, 726b2–11, 727a3–4, 727a31–727b5, 728a26–30.

female, being incapable of concocting (πέττειν) the nutriment in its last stage into semen (σπέρμα) (and this is either blood or that which is analogous to it in animals which are bloodless) owing to the coldness (ψυχρότητα) of her nature ... For the menstrual blood (τὰ καταμήνια) is semen (σπέρμα) not in a pure state but in need of working up (δεόμενον ἐργασίας), just as in the formation of fruits the nutriment is present, when it is not yet sifted thoroughly, but needs working up to purify it. Thus the menstrual blood causes generation by mixture (μιγνυμένη) with the semen (τῇ γονῇ), as this impure nutriment in plants is nutritious when mixed with pure nutriment.[2]

Aristotle argues against the Hippocratics that the male and female contribution develop from the same source of nutriment, which becomes blood. While Aristotle points to a common root of the male and female contribution, this passage has all the indicators of the reading that woman is the material whose being is for the sake of form. Woman is described as an impotent male, defined by incapacity, impure and in need of the work of semen. Women's incapacity is due to her coldness. This chapter investigates the details of how blood as a homoeomerous part in the body comes to have its nature from the powers at work in the elemental forces of hot and cold and moist and dry and how that blood becomes semen through that work of hot on moisture in the process of concoction. Considering these details will show first, that material in Aristotle at its most basic level is powerful and already imbued with specific potentiality and second, that the male contribution develops from the same process as the female contribution though it reaches a higher capacity by achieving a degree of vital heat. The case that material has specific power and that the distinction between the male and female contribution – the form and matter – depends on a process they both share, suggests that the distinction between form and matter is less obvious than is generally considered. This distinction is not between what defines and what is stuff, but between a potentiality in material form and a potentiality in the most basic being of matter – elemental matter – that does not depend on form. Semen, which is the tool of soul in the sense of what manifests it, does its formal work through the material power of heat that comprises it. Menses, which is potentially alive, is a potential 'acted upon' that is like what acts upon it. This way of construing the difference between form and matter invites us to rethink how the material of semen is the tool of form as that whereby it is put to work in the world.

By contrast to the view Aeschylus puts in the mouth of Apollo, Aristotle affirms a positive contribution from both the female and the male – the menses

---

[2] *GA* 728a16–21, 26–30.

and the semen. In *GA* II.4, Aristotle defines semen, like menses, as a 'useful residue', by contrast to those who think it is a waste product.³ These materials are what is left when nutriment is concocted into blood. It is because semen and menses are residue that people with more fat are less fertile: the nutriment has been concocted into fat with little residue left over for generation.⁴ Like menses, semen is blood concocted to a certain degree, a degree which makes it capable of bringing forth life. Menstrual blood is incomplete semen in which the nutriment is present but in need of working up. This final working up makes of the residue that which has the power to produce life. The fully worked up residue can produce life and the residue that is not quite fully worked up needs to be acted upon by the fully worked up residue for it to become life.

The difference is not only a difference within a body of what the male principle and what the female principle do in that body, but also the difference between the body that will be able to act as a male principle and the body that acts as a female principle.

> But the male and female are distinguished by a certain capacity (δυνάμει) and incapacity (ἀδυναμίᾳ). (For the male is that which can concoct (δυνάμενον πέττειν) and form and discharge a semen (σπέρμα) carrying with it the principle (τὴν ἀρχὴν) of form (τοῦ εἴδους) – by 'principle (ἀρχὴν)' I do not mean a material principle (ὕλης) out of which comes into being an offspring resembling the parent, but I mean the first moving cause (κινοῦσαν πρώτην), whether it have power to act as such in the thing itself or in something else – but the female (θῆλυ) is that which receives (δεχόμενον) ~~semen~~, but cannot form (συνιστάναι) it or discharge (ἐκκρίνειν) it.) And concoction works by means of heat (θερμῷ). Therefore the males of animals must needs be hotter (θερμότερα) than the females. For it is by reason of cold (ψυχρότητα) and incapacity (ἀδθναμίαν) that the female is more abundant in blood in certain parts (τόπους τινὰς) ~~of her anatomy~~, and this abundance is an evidence of the exact opposite of what some suppose, thinking that the female is hotter (θερμότερον) than the male for this reason, i.e. the discharge of menstrual fluids (καταμηνίων πρόεσιν).⁵

In this passage, which appears much later in *Generation of Animals* in the context of how sex differentiation occurs in offspring, Aristotle explicitly

---

³ *GA* 724b34–725a2, 725a12–34, 738a31–6.
⁴ *GA* 726a4–5, 727a34–727b1. Note that this makes the material for generation the very last residue of the body. When Aristotle says that nature is a good householder and so uses the best material for the best parts, it seems that generation of offspring is quite far down the scale.
⁵ *GA* 765b9–20.

associates male with heat and female with coldness in a way that associates male with a capacity for life and female with incapacity. Coldness is connected to the abundance of moisture – blood – in the female, following this similar association of male with fire and female with water or moisture that is found in the Hippocratics. Aristotle distinguishes the male and female by their capacity or incapacity to heat and therefore, to form. Different levels of concoction of the useful residue lead some residue – sperm – to become capable of bringing life into other residue – menses – which can be worked up to a certain degree of thickness. This first residue so concocted affects the second residue in a way that forms something new that is distinct from either residue. The concoction that forms semen out of residue and imparts the power to concoct into the semen occurs in the semen through a certain kind of heat, a heat that comes only from other things that share this heat.

This chapter focuses on Aristotle's account of the elements and the elemental forces to explain how heat works in generation to bring forth sexually differentiated life. The argument will support the view that at its most micro level, φύσις is an internal source of movement, and that the form and material that constitute the new substance are more intertwined than we tend to understand in Aristotle's metaphysics. This chapter addresses how the female contribution is a positive one, not merely a lack, and the male contribution is rooted in the material work of vital heat as it arises out of elemental forces. Material has its own capacity. Form's work is in and through material.

Concoction as Mastery of the Moisture

The material difference between the female and the male contribution seems to be a difference in degrees of heat. Yet describing this distinction in material terms does not yet overcome the normative hierarchical metaphysics involved. The work of concocting blood into semen occurs when the heat gains a certain control over the moisture on which it works. It is hardly deniable that Aristotle uses language of domination in this process. He explains the work of both heat and cold as active powers at the beginning of *Meteorology* IV, 'When the hot and cold are masters of the matter they generate a thing.'[6] The claims of this section are: that the hot and cold acting as form to master the moisture in concoction are acting through their material power, that the mastering role shifts between the passive and active contraries, and that what is acted upon positively affects the process. Together these claims contribute to the view

---

[6] *Meteor.* 379a1.

that form works materially, in this instance at the most elemental level, and that material can play what is traditionally conceived as a formal role.

Having begun by dividing this work of mastering what is worked on between hot and cold, Aristotle continues to describe the process of concoction as:

> a process in which the natural and proper heat [τοῦ φυσικοῦ καὶ οἰκείου θερμοῦ] of an object perfects the corresponding passive qualities [the dry and moist], which are the proper matter [ἡ οἰκεία ἑκάστῳ ὕλη] of any given object.[7]

Aristotle divides the active work of the elemental forces of hot and cold from the passive work of the elemental forces of dry and moist. As the treatment of elemental change in Chapter Three evidences, Aristotle follows the Hippocratics against those among his contemporaries who consider elements the first level of composition when he maintains that earth, air, water and fire are composed of elementary forces.[8] In *Parts of Animals*, Aristotle explains that the elementary forces of wet and dry, hot and cold together 'form the material of all composite bodies' and produce the various differences in the bodies, 'heaviness or lightness, density or rarity, roughness or smoothness, and any other properties of bodies as there may be'.[9] The elements are composed of more basic powers, powers with distinct capacities that exist in mixtures. These forces work toward the next level of composition, which is homogeneous parts – bone and flesh, and so forth – which compose the heterogeneous parts – the face, and the hand and other organs. Aristotle writes that 'it appears manifest', indicating general acknowledgement of those who

---

[7] *Meteor.* 379b18-20. See *Meteor.* 388a21-4, 'The homogeneous bodies, it is true, are constituted by a different cause, but the matter of which they are composed is the dry and the moist, that is, water and earth (for these bodies exhibit those qualities most clearly). The agents are the hot and the cold; for they constitute and make concrete the homogeneous bodies out of earth and water.'

[8] Aristotle is able to maintain that all natural substances are composites of a sort because the elemental forces that compose the elements do not exist distinctly from another force. Aristotle seems to contradict this view when he defines an element in *De Caelo* as 'a body into which other bodies may be analysed, present in them potentially or in actuality (which of these, is still disputable), and not itself divisible into bodies different in form. That, or something like it, is what all men in every case mean by element', *De Caelo* 302a15-19. Aristotle continues in *De Caelo* to draw conditional claims about whether the elements are the most basic indivisible material parts (303a25-7, 304b25-7), and he concludes that elements are generated from elements, which is consistent with coming from elemental forces, because the previous composite of the elemental forces is an element (305a31-3).

[9] *PA* 646a18-21.

consider the matter, 'that these are the properties [hot or cold, dry or moist] on which even life and death are largely dependent, and that they are moreover the causes of sleeping and waking, of maturity and old age, of health and disease'. Thus, Aristotle concludes, 'For hot and cold, dry and moist, as we stated in a former treatise, are the principles (ἀρχαὶ) of the natural elements.'[10]

In some places Aristotle refers to the female as colder and in other places as moister, which would make the female both the contrary of heat and the contrary of dry, both an active and passive principle. This distinction is confirmed by Aristotle's account of concoction. The passive principle in concoction is the 'proper matter' which is perfected by heat, an active principle. Aristotle continues that 'the primary source (ἀρχὴ) [of concoction] is the proper heat (θερμότης) of the body'. Proper heat is the starting point of concoction. The end of concoction can be the nature of a thing, as Aristotle explains, 'nature, that is, in the sense of the form and essence'.[11] Having associated the moist with matter and heat with form, Aristotle concludes:

> Concoction ensues whenever the matter, the moisture, is mastered (κρατηθῇ). For the matter is what is determined by the natural heat in the object, and as long as the ratio between them exists in it a thing maintains its nature.[12]

The matter that is mastered is another elemental force – moisture – which heat works upon. Concoction's work of heat on moisture would seem to confirm that the masculine and the feminine are distinguished along the lines of the active and passive, the formal and material, even at the level of elemental forces. Yet while the language of mastery suggests that there is a fixed position of the elemental force whose role is to master, Aristotle's discussion of the coming-to-be and passing-away of elemental forces demonstrates that different material forces play different roles in different contexts.[13]

---

[10] *PA* 648b4–11.
[11] *Meteor.* 379b24–6.
[12] *Meteor.* 379b33–5. See *Magna Moralia* 1210a15–21, where Aristotle compares utility friendship to the relation between fire and water: 'For if you like to set down fire and water as the extreme opposites, these are useful to one another. For fire, they say, if it has not moisture, perishes, as this provides it with a kind of nutriment, but that to such an extent as it can get the better of; for if you make the moisture too great, it will obtain the mastery, and will cause the fire to go out, but if you supply it in moderation, it will be of service to it.'
[13] Eric Lewis similarly argues that in elemental change, one elemental force is form and another material. His view makes the active forces always on the side of form and the passive forces always on the side of material. Lewis explains that this work of mastery of the active contraries on the passive contraries as organising and informing it in a way that makes it more complex. He argues that neither force overcomes the other so that they maintain their individual identity even in unity in Eric Lewis, 'Introduction', p. 23.

Eric Lewis argues from Alexander of Aphrodisias's commentary on Aristotle's *Meteorology* that Aristotle distinguishes between uses of συνιστάναι in *Meteorology* as determining, organising, condensing and concocting in order to maintain a hierarchy within the elements, where the more condensed a thing, the more organised it is, as in the shift from air to water to earth. Lewis connects the condensing work of συνιστάναι to the defining work of ὁρίζειν, which both defines and determines, placing fire at the top and moisture on the bottom. As Lewis puts it,

> The alteration of the organization of something will therefore alter its capacities. In other words, if to gain organization is, among other things, to gain capacities, and there is an organization (and so a set of capacities) which something has when it is most fully what it is, then it is fair to say that something is determined by having (or gaining) a certain capacity, and is therefore also defined by this capacity. To be determined would be to take on an organization such that one gains a characteristic capacity or set of capacities. To have these capacities would be to exemplify the form of this thing, to be this thing according to its definition.[14]

Lewis allows that these functions and powers that characterise a thing can be either passive or active, the capacity to be affected or to affect, but he seems to further an Anaxagorean or Empedoclean view of material when he argues that the different degrees of organisation give material stuff different degrees of ὁρίζειν, or definition and determination. Lewis's account offers a material account of form, based on processes of concoction, but his account leads to a view of form in terms of increased density. Furthermore, Aristotle shows in his account of elemental change how each elemental force can determine and be determined in the process of the change of elements, which would suggest that rarefying as much as condensing on Lewis's account would have to do the work of form.

In elemental change, the passive contraries can be active and changing, while the active contraries can be the ones that are passive and enduring through the change, as when fire changes into air, and the heat remains, but the change is from dry to moist. Moreover, the passive contraries are not themselves privative or without capacities of their own. The passive contraries are the contraries whose capacity is to be worked upon in particular ways, which is not a privation of capacity but a particular way of having a potential, one which shows the likeness between contraries that makes the active work on the passive to actualise

---

[14] Lewis, 'Introduction', pp. 32-4.

its capacity to be like the active. Alexander explains Aristotle's claim that the 'moist and dry are passive, for it is in virtue of its being acted upon in a certain way that a thing is said to be easy to determine or difficult to determine', by saying that the moister elements are undetermined because they have no limit of their own.[15] While Alexander finds the passive contraries to have no limit of their own, even fire, which we generally associate with the active contrary of heat, is said to be without limit and so incapable of causing soul in *De Anima*. If having no limit of their own is the distinguishing factor of the active and passive contraries, then the chief active contrary would be passive, having no limit of its own. Alexander explains that generation of plants and animals occurs then when these passive and undetermined powers that characterise the moist elements – water and air – are 'affected and changed by the active' powers such as hot and cold. For Alexander, passive qualities are the natural matter of all things.[16] Interestingly, Aristotle calls passive forces of moist and dry the principles – ἀρχαί – of physical bodies.[17] These principles are characterised as passive, παθητικαί, but this does not mean without their own character. Moisture can be material to heat's form, but moisture itself is also 'vivifying'.[18] Even as the different forces have different roles in elemental change, the two pairs of contraries in elemental change seem capable of exchanging positions.

The contraries have more pronounced and solidified positions as active and passive when they characterise the material that works or is worked upon in the formation of homoeomerous parts. The focus of *Meteorology* IV is on how the various processes of the active principles work on the passive principles, which comprise the elements or simple bodies. In *Meteorology* IV.2, Aristotle writes that we have to consider how these δυνάμεις or powers affect 'already constituted natural bodies (τῶν φύσει συνεστώτων ἤδη) as the substrate, ὑποκειμένων'.[19] In considering how food becomes blood and blood becomes flesh, Aristotle explains that while moisture may be without its own limit (moisture coupled with cold – water – is the only thing that cannot be thickened or concocted at all),[20] moisture is characterised differently depending on

---

[15] *Meteor.* 378b21–5, 381b29. Alexander of Aphrodisias, *On Aristotle's* Meteorology 4, 180, 10–21, 198, 28–30.

[16] Alexander of Aphrodisias, *On Aristotle's* Meteorology 4, 182, 14–15, 185, 13–15. *Meteor.* 380a9.

[17] *Meteor.* 381b24–5. Both H. D. P. Lee in *Meteorologica* and E. W. Webster in *The Complete Works of Aristotle* translate the Greek for principle (αἱ ἀρχαί) as elements in this passage.

[18] *GA* 733a11. See *Juv.* 466a20–4.

[19] *Meteor.* 379b11. Both Lee and Webster translate ὑποκειμένων as material and Lee translates δυνάμεις as properties, Webster as qualities and Alexander as forms.

[20] *Meteor.* 380a34.

how it mixes with dry. The material is characterised because of the way that it can be affected and it can be affected differently depending on how it is mixed with other forces that can be affected.

Though the elemental forces do not have their own capacity to limit themselves, both active and passive contraries contribute to the character of what they form. The active contrary is limited by the passive contrary on which it works, which Aristotle explains with reference to the way heat produces something different depending on what it determines. Much of *Meteorology* considers the differences in the various mixtures of the dry and the moist of boils, phlegm, milk, oil, wine, urine, vinegar, lye, whey, honey, blood, natron, horn, pus, rheum, excretion and pottery.[21] Aristotle points to how each of the elemental forces – the hot and the cold and the moist and the dry – are at work in the process of distinguishing what is formed, arguing that it is both 'their capacities of action' and their 'characteristic affections which express their aptitude to be affected' that define a thing. 'These affections differentiate most bodies from one another', Aristotle explains before going on to enumerate the specific ways that hot and cold work on differently organised mixtures of the moist and dry.[22] Ripening is dry heat, even natural heat, working from an airy state into a watery state. Something is raw when it lacks the natural heat because the moisture is too great. Boiling is moist heat working on airy or watery material. Broiling is an external dry heat working on a moist or airy matter, which leads to different results depending on the proportion of air and water and earth in the matter. Parboiling is the work of cold or lack of heat on moist or airy material, and so forth.[23] In each of these cases, a particular way that heat and cold are related to moist and dry works on a particular compound of moist and dry. All the different ways that the passive forces can be construed lead to different results of these processes of hot and cold working on them. These middle chapters of *Meteorology* IV point us to the way that passivity is not incapacity but a specific capacity to be affected.

In *Parts of Animals*, Aristotle entertains the possibility that fire itself may be a case of a thing that though hot when it burns is not itself hot:

> There is no knowing but that even fire may be another of these cases [in which the substratum owes its heat to an external influence]. For the substratum of fire may be smoke or charcoal, and though the former of these is always hot, smoke being an uprising vapour, yet the latter becomes cold when it is

---

[21] *Meteor.* 380a21–3, 380b31–4, 381a7, 384a4, 12–17, 21–4, 384a35–384b1, 380b5.
[22] *Meteor.* 385a2–6, 19–20.
[23] *Meteor.* 380a20–6, 32, 380b12–23, 381a11, 381a24.

extinguished, as also would oil and pinewood under similar circumstances. But even substances that have been burnt nearly all possess some heat, cinders, for example, and ashes, the waste-products of animals, and, among the excretions, bile; because some residue of heat has been left in them after their combustion. It is in another sense that pinewood and fat substances are hot; namely, because they rapidly assume the actuality of fire.[24]

That which fire works on can affect the heat in fire. The reason is that the substratum of fire – not the contrary that remains the same when it changes to another element, but the natural composite in which fire can inhere – can become cold, by contrast to other substrata which remain hot even when they have been burnt. It is only those substrata that have internal heat and are hot *per se*. Aristotle seems to be speaking about composites that have internal heat, as something like blood or another homogeneous substance that is a mixture of elements. The residue of heat that remains points to a heat internal to the substratum that makes it hot *per se*. A third way that substrata are hot in relation to fire is that they quickly 'assume the actuality of fire', that is, they 'catch fire', as we say, easily. Some substrata have the capacity to burn, but they may not have an internal source of heat. Fire above the level of the elemental depends on whether it inheres in something dryer or wetter, airier or earthier.

Every natural part forming a natural substance has some active force and some passive force that characterises it depending on what is becoming what. The concoction of blood is the work through which – at a certain degree – life comes into a heated moisture; successful concoction or heating of this moisture will change it from being one kind of moisture (blood) to another that is mixed with more air (semen). Semen will be able to work again on blood to form animated breath. Heat works here, but so do the other elemental forces – cold, moisture and dryness. When the concoction is successful and becomes that which can produce life and does so, the living thing will also need cold and moisture to endure.

Cold and moisture are associated with the female body and the female residue, and yet cold is a positive capacity, not merely the absence of heat, though it was generally considered privative among Aristotle's contemporaries. The two couplets are contraries, but not contraries as the presence or absence of some power.[25] For example, in the case of blood, which is what is concocted in

---

[24] *PA* 649a22–9.
[25] *GC* 324a11.

generation, the substratum as Aristotle describes it in *Parts of Animals* is not itself hot but becomes hot from something external, which leads Aristotle to conclude, 'it is plain that cold is not a mere privation, but a fact of nature'.[26] The contrary of heat, which in other contexts appears as a privation, στέρησις, is here displaying its own power; at the same time, the contrary of heat as the substratum of heat, moisture, displays its own power.

The non-privative contrary relationship between cold and hot is evident in Aristotle's view that coldness is more suitable to sensation and intelligence and hotness to strength. He even goes so far as to say in *Parts of Animals* that 'of sanguineous animals, those are the most intelligent whose blood is thin and cold. Best of all are those whose blood is hot, and at the same time thin and clear. For such are suited alike for the development of courage and of intelligence.'[27] Such animals would have to be hot and cold, since thinness is a result of cooling while thickness is a result of heat. Hot and cold balance one another out to the extent that the hotter things, which are most alive and intelligent, are both hot and cold, the brains being characterised as cold.[28] Similarly, in *On Youth, Old Age, Life and Death* and in *On Respiration*, Aristotle explains that nutrition depends on 'the natural fire', which requires cooling in order not to be exhausted and thus destroyed.[29] Aristotle gives place and gender to these conditions when he associates the upper parts, male and the right side to what is superior.[30] While there is a hierarchy here, the cold has a capacity and a role in the body of its own.

This analysis recognises that what is mastered, the moisture, has a particular power to variously respond to what acts upon it. In *Meteorology* IV.5, Aristotle names two causes other than material, τὰ αἴτια τὰ παρὰ τὴν ὕλην δύο – the moving or efficient cause, τό τε ποιοῦν, and the quality or the affection, τὸ πάθος.[31] Aristotle continues that the agent is the efficient cause and the quality is the formal cause (τὸ δὲ πάθος ὡς εἶδος). The agent works through the qualities as the efficient cause works through form. This view does not make the form the tool of the efficient cause but the active means by which it works. The moving cause acts through the hot and cold, which are productive of change, where hot is more of a productive agent and cold

---

[26] *PA* 649a20-1. Despite Aristotle's claim at *GA* 743a37 that cooling is mere deprivation of heat.
[27] *PA* 648a8-11.
[28] *PA* 652a35-6.
[29] *Respiration* 474b10-12; see also *Juv.* 470a5-6, 20-5, 476a15-18, 479a8-9.
[30] *PA* 648a11-14.
[31] *Meteor.* 382a28-30. Lee and Webster both translate πάθος as quality.

is a destructive one. The hot can work in different ways – by coagulating or melting, ripening, broiling or boiling – depending on what it is working on and whether the hot thing is hot in virtue of itself or merely as an attribute, where it could remain itself and not be hot. While heat is associated with form and moisture with matter, the process whereby the form is at work is material both in the sense that heat itself is material and in the sense that the material of semen remains moist, but its moisture is transformed when characterised as hot in virtue of itself and not accidentally, as blood is.

Internalised Heat: From Blood to Semen

Aristotle describes concoction as the process whereby heat masters the moisture. The analysis of the ways that heat works shows that heat is not form *simpliciter* for Aristotle, but form specifically in concoction. Aristotle draws a distinction between a heat that characterises homogeneous substances like blood without themselves making the blood hot and the heat that is internal to such a substance, characterising it *per se*. The heat at work in blood does not come from the blood but from external forces on it – the movement in the heart. In *Parts of Animals*, Aristotle attributes the disputes over what is hotter to the fact that things are called hot or hotter in several ways.[32] One body is hotter than another if it imparts more heat to an object in contact with it, in another sense if it 'causes the keener sensation when touched', especially if the touch causes pain, in another if one melts a substance or sets it on fire more easily than another, in another the hotter is larger than the cooler, in yet another the body that takes longer to cool or quicker to heat is considered hotter. While a body cannot be hot in all of these ways at once, some of these characterise the hot *per se* and some characterise the hot *per accidens*. The hot *per se* is hot in its own right, yet Aristotle notes that it is 'no simple matter' to decide which is hotter. It is not just difficult to tell which is hotter, but to tell whether a thing itself is hot, 'For the actual substratum (ὑποκείμενον) may not itself be hot, but may be hot when coupled with heat as an attribute.'[33] Heat seems to be an attribute in the case of blood. Aristotle uses 'actual substratum' as that which we look to in order to determine what a thing is, and the external source as a mere attribute, the account of substratum that Aristotle associates with substance in *Metaphysics* Z.3. Substratum in this case is what is definitive of what the thing is, the substratum as substance that is a separable τόδε τι. The heat that makes the substratum itself hot would be the

---

[32] PA 648a37. See *On Breath* 485a28–485b8.
[33] PA 649a16–17.

form that constitutes the substance as hot, while the external heat would be an accidental attribute.

In *Parts of Animals* II.5, Aristotle notes that blood is one of these things that is hot in one way but not definitively:

> But the substratum of blood, that which it is while it is blood is not hot. Blood in a certain sense is essentially (καθ' αὑτὸ) hot, and in another sense not so. For heat is included in the definition (τῷ λόγῳ) of blood, just as whiteness is included in the definition of a white man; but so far as blood becomes hot from some external influence (κατὰ πάθος), it is not hot essentially (οὐ καθ' αὑτό θερμόν).[34]

Blood is a substratum that is heated without producing the source of heat, while semen becomes what it is – hot – through concoction to the point where it becomes capable of producing this heat on its own. Semen's capacity to become hot in this way will be notably shown to depend on the kind of material, the useful residue, that is heated. Semen becomes capable of carrying the heat beyond the body in which it originates to become the source of soul in another – not only by heating but by heating to the point where what is heated has the capacity to heat in another. Semen's capacity to heat in another stands in contrast to blood, which remains hot and moist only as long as it remains in the living body.

Aristotle discusses the dry and the moist in the same chapter of *Parts of Animals* where he explains that blood is in a sense essentially hot, but in another sense not so. Blood is compared to ice, which is actually and accidentally dry but potentially and essentially wet.[35] This is a peculiar configuration because in the case of ice, the potential is what a thing most essentially is. Ice is dry because it is excessively cold, yet it remains potentially wet as water. Aristotle explains that the living body is what makes some substances hot and moist to argue that the being of the blood depends on the living body remaining alive.[36]

Aristotle's view of concoction as a process of heat mastering the moisture coupled with his account of the transition from blood to semen indicates that concoction occurs when heat has become most fully internalised in relation to moisture. The moisture has been transformed, while still characterising what is concocted as moist. The Hippocratic author distinguished the power of fire and

---

[34] *PA* 649b24-7.
[35] *PA* 649b12-13.
[36] *PA* 649b28-32.

of water by saying that fire can move all things and water can nourish them, and Aristotle appears to follow this distinction.[37] In the work of concoction, the heat of fire comes to characterise the moisture of blood in such a way that it is capable of bringing forth that heat in another. The moisture is mastered not by ceasing to be moisture, but by that which masters characterising it.

As the previous section establishes, while Aristotle maintains that heat drives change in living things because it seems like the active constructive elemental force, even fire depends reciprocally on moisture which has its own distinct capacities. In *On Youth and Old Age, Life and Death*, Aristotle opposes fire and water as 'reciprocal causes of generation and decay', which 'do not possess identical powers'.[38] Aristotle rejects the view that the sun is nourished by moisture, a view which would seem to make celestial bodies dependent on sublunar ones, but he acknowledges that fire is fed by moisture. He ascribes a particular role to fire in natural things. In *Meteorology*, he explains that everything decays but fire, 'for earth, water and air all decay, since all are the matter of fire (ὕλη τῷ πυρί)'.[39] This passage suggests that fire, the element comprised of the elemental forces hot and dry, endures but the combinations that include cold or moisture come to be and pass away. Destruction is what happens when 'what is being determined (τοῦ ὁρίζοντος) gets the better (κρατῇ) of what is determining it (τὸ ὁριζόμενον)'.[40]

In the destruction of the living body, heat is not overtaken altogether, but an external heat overtakes the body's natural heat.[41] If what is determined in concoction is moisture and what is determining is heat, in destruction (σῆψις) it is not just that the moisture overtakes heat, but that the natural heat (κατὰ φύσιν θερμότητος) of the moist body is destroyed by an external heat (ἀλλοτρίας θερμότητος). When Aristotle follows up by explaining that living things become dryer in decay 'for as its own heat (τοῦ οἰκείου θερμοῦ) leaves it its natural moisture (κατὰ φύσιν ὑγρόν) evaporates, and there is nothing to suck moisture into it (this being the function of its own heat (ἡ οἰκεία θερμότης), which attracts and draws moisture in)',[42] he points to how in the living body, the moisture that characterises it depends on the work of natural heat.

---

[37] *Vict.* I.3.
[38] *Juv.* 465a14–16. See *Meteor.* 355a5.
[39] *Meteor.* 379a15–16, my translation. See also *De An.* 416a26–9.
[40] *Meteor.* 379a11–12, Lee's transation.
[41] *Meteor.* 379a16–17.
[42] *Meteor* 379a24–6. See *De An.* 416a27–9. 'And of those that are, not all seem to be nutriment in the same way for each other, but water is nourishment for fire whereas fire does not nourish water. It seems to be most of all the case in the simple elemental bodies that one nourishes and the other is nourished.' See *GA* 784b6 where Aristotle maintains that all decay is caused by heat but not by innate heat.

A living body has natural moisture as well as natural heat. The natural moisture depends on the natural heat to draw the moisture into the body. Decay occurs when the moisture ceases to be determined by the heat but begins to determine it. At the same time, the natural heat depends on the persistence of the natural moisture, which when it dries up, allows the natural heat to be overtaken by external heat. The body loses the moisture when the heat stops pulling the moisture in. Concoction occurs when the heat masters the proper matter, moisture. As this passage makes clear, that mastering does not mean that the moisture disappears or ceases to be moisture, but that it is determined in a way that maintains the ratio between heat and moisture, between the moving and the nourishing.[43]

What underlies blood is not heat because the heat is not καθ' αὑτό, according to itself, but what blood undergoes or suffers. While blood's heat seems more essential to blood being what it is than the whiteness of man is essential to it, as Aristotle compares the blood's heat to the man's paleness, both are external. Heat in blood is like whiteness, τὸ λευκόν, a kind of pallor that is not essential to a person, but brought on by some sickness or shock, externally produced. Evidence that supports the accidental nature of blood's heat is that blood loses both its heat and its moistness when separated from the body, while other substances, like yellow bile, become moist when separated from the body.[44] The living body is not itself essentially moist or dry, but hot, and it is this heat which keeps blood moist. Some heat in the body is the cause of the hotness of its parts; the hotness of the parts is not the source of its heat. Against those who might think that blood is the part that is the source of life and the heat that brings this life, Aristotle argues that blood itself has some other source of heat. This other source will turn out to be the natural heat of the body, the same heat that will then build heat up more in order to concoct it into semen.

> Now since everything that grows must take nourishment, and nutriment in all cases consists of moist and dry substances, and since it is by the force of heat that these are concocted and changed, it follows that all living things,

---

[43] Aristotle seems to draw directly from the Hippocratic author who writes, 'Fire has the hot and dry, water the cold and moist. Mutually too fire has the moist from water, for in fire there is moisture, and water has the dry from fire, for there is dryness in water also', *Vict.* I.4. As Hynek Bartoš notes, this configuration makes the elements for the Hippocratic author a mixture of three and not two elemental powers, since both fire and water have the moist and the dry, heat and the cold are what ultimately distinguish them, in 'Aristotle and the Hippocratic *De Victu* on Innate Heat and the Kindled Soul', p. 295.

[44] *PA* 649b29–35.

animals and plants alike, must on this account, if on no other, have a natural source of heat . . .[45]

Blood is the fundamental homogeneous part, and it relies on heat to be moist. While blood's heat is heat *per accidens*, in concoction blood takes on the natural internal heat of the body of the father in order to become semen. That is to say, blood becomes semen when it becomes that which has heat καθ' αὑτό. Concoction, which transforms blood into semen, occurs when the moisture is fully mastered by the heat. The heat that is external to the blood becomes internalised in semen in such a way that emanating heat itself, it is able to bring heat into another thing so that it might then emanate heat in such a way that it is capable of causing another thing to emanate heat itself. Concoction is the perfection of contrary passivities by means of natural heat. Alexander notes that Aristotle points to the work of concoction as occurring out of (ἐκ) contrary passivities rather than out of (ἐκ) underlying matter, but Aristotle proceeds to call the passive contraries the 'matter proper to the particular thing, ἡ οἰκεία ἑκάστῳ ὕλη'.[46] The matter – the contraries of moist and dry – is passive because it is perfected by becoming affected. It is in this sense that the moisture is mastered by the hot in concoction. The mastery is not of what is independent and self-regulating against what is not, but of what acts in particular ways on what is capable of being acted on in particular ways. Both are required, both work in specific and not just generic or structural ways of acting and being acted upon, as the details of the middle chapters of *Meteorology* IV attest. For example, when there is more water than earth the mixture is thickened by fire. When there is more earth than water, the mixture is solidified by fire. When air is mixed in and oil is the mixture, it is thickened by cold and hot.[47] This process occurs through the work of breath. Through breath, semen produces heat on its own accord, rather than being dependent on the living body in which it inheres to be hot. The independence of semen's heat does not make semen or its force immaterial.

The difference between the male contribution and the female contribution, between what serves as form and what serves as material, is the difference between what is its own source of heat and what relies on heat in something outside of it. What relies on heat in something outside of it will be transformed in generation to a natural substance that produces its own heat. In

---

[45] PA 650a2–6.
[46] Alexander, *On Aristotle's* Meteorology 4, 186, 19–21, *Meteor.* 379b20.
[47] *Meteor.* 383a19–20, 388a31–3, 388b9–10, Alexander 208, 21.

order for the source of heat to work on what has an external source of heat, the latter must be capable of becoming hot. Menses as even more concocted than blood shares the potential of hot with semen. It is acted upon as that which is like rather than unlike.

Material and Teleology in Aristotle's *Meteorology*

Before turning to the account of how semen's power is formal and material, the very possibility that semen can operate on the basis of material must be defended with a return to the question of how much work material can do. Those who argue that the male contribution cannot be grounded in material interactions point to the last chapter of Aristotle's *Meteorology* to argue that while uniform or homoeomerous parts are caused by the elements or elemental forces, nonuniform or nonhomoeomerous parts require further causes. In *Meteorology* IV.12, Aristotle writes:

> Now heat and cold and the motions they set up as the bodies are solidified by the hot and the cold are sufficient to form all such parts as are the homogeneous bodies, flesh, bone, hair, sinew, and the rest. For they are all of them differentiated by the various qualities enumerated above, tension, ductility, fragmentability, hardness, softness, and the rest of them: all of which are derived from the hot and the cold and the mixture of their motions. But no one would go as far as to consider them sufficient in the case of the non-homogeneous parts (like the head, the hand, or the foot) which these homogeneous parts go to make up. Cold and heat and their motion would be admitted to account for the formation of copper and silver, but not for that of a saw, a bowl, or a box. So here, save that in the examples given the cause is art, but in the non-homogeneous bodies nature or some other cause.
>
> Since then, we know to what class each of the homogeneous bodies belongs, we must now find the definition of each of them, i.e. what is blood, flesh, semen, and the rest? For we know the cause of a thing and its definition when we know its matter and its definition – and best when we know both the material and the formed factors of its generation and destruction, and also the source of the origin of its motion.[48]

Semen occupies a liminal position in the discussion of the homogeneous and the nonhomogeneous parts since it is a homogeneous part that is the source

---

[48] *Meteor.* 390b3–19. See also *GA* 743a4–743b24.

of change that produces nonhomogeneous parts. On the one hand, semen is formed out of the elements, as a homogeneous part. On the other hand, semen as the source of soul must be caused by what has soul, which we associate with the moving, not material, cause. The argument is that material cannot cause the purposiveness of form, and hence of semen, but only how it must act of necessity according to its material composition in relation to other elements or elemental forces. The restricted material view rejects the possibility that material composition of the semen can explain how it animates. Necessity can explain why the semen must be composed of the elemental forces that compose it to enable it to act as it does, but necessity of the elemental forces cannot explain the purpose for the sake of which semen acts as it does.

On this view, semen's capacity to animate requires that it be composed of certain material, but the material does not introduce the capacity for this function; it enables it. The semen is composed of this material by some being – the father – who uses semen in order to fulfil his end. The material contributes to this cause, but is not sufficient for fulfilling this end. The material is determined by the form in the sense that semen is made of certain material at the behest of form so that semen might do its work.

Yet as that which is the site of the joining of material and formal power, semen's material manifests this command of form. Semen's material is determined by the form of the father, but semen's material is also what determines form in the menses. Actual form, in the being of the father, requires the materiality of semen to do its work. Potential form, in the being of the embryo, becomes soul through the material capacities of semen. A process of heating produces a part that internalises the heat in a way that gives it certain animating power. The useful residue that is semen becomes concocted so that it can bring forth life by the formal powers of the father.

Against the notion that generation is alteration or change in the arrangement of materials, Aristotle argues that generation does not occur through a haphazard arrangement of stuff, but through the generation of particular forms from previous enmattered substances of similar form. This form is actualised in that which is potentially alive. The aim toward actualising what is alive makes natural generation regular and consistent. Gill argues that this view does not negate the goal-directedness of the elements, whose dual function as elements is to move toward a particular place and to form nonhomoeomerous parts.[49] In *Physics* II.1, Aristotle includes elements among those beings which are natural,

---

[49] Mary Louise Gill, 'The Limits of Teleology'.

having an internal source of movement.⁵⁰ Gill argues against a view that takes *Meteorology* IV.12 to mean that all of nature must be explained through formal or final causes. Teleological claims explain elements only insofar as they contribute to living substances or artefacts, but uniform parts and elements can be formed merely out of necessity and not for some more complex end. The language of necessity views elements from the perspective of that which they come to form and not on their own terms. But elements themselves have an end, which is to be in a particular place and they appear to strive toward that end unless impeded by something external.

Meteorology IV makes the argument for how all the parts of an organism can be explained from their material and from the form of the organism.⁵¹ The difficulty to which Aristotle is pointing in the last chapter of the book is that farther down the teleological line from the whole organism it is less clear why the material is what it is for the sake of the whole organism. Affirming that all things are what they are by their function, Aristotle notes this is also true of fire but it is not as clear what fire's function is, just as copper and silver have not-easily-perceivable functions.

This tension between necessity and teleology of elements is the tension found in material itself. Matter is both characterised on its own terms, as the elements are, and characterised by form as the body and functional matter of natural living substances. The ways that fire itself functions as form have been established in ways that show that material has a character independent of form. Fire functions as form for natural substance, which transforms fire into vital heat. Fire on this view becomes limited by the work it is doing for soul, as the tool of soul. Fire can be both powerful on its own account and working on behalf of natural form. The last chapter of *Meteorology* IV establishes that some parts can come into being outside of a natural organism, which makes possible 'an ineliminable "bottom up" story ... not only about inanimate material objects but about living organisms as well'.⁵²

Cold and heat can do the work of forming at the level of the homogeneous, but it still seems that form beyond what cold and heat can do is needed to

---

⁵⁰ *Phys.* 192a9-15. Helen Lang argues that this movement is toward a particular place in *The Order of Nature in Aristotle's Physics*, yet the internal principle toward movement seems to be just as much a capacity to transform that with which they interact, for example, fire to heat, water to moisten. Gill argues in *Aristotle on Substance* that elements are not opposed because they are just material and so form is not striving against the material nature, pp. 235-6.

⁵¹ *Meteor.* 389b28-9.

⁵² Gill, 'The Limits of Teleology', pp. 344, 348.

shape the nonhomogeneous parts. Yet this work of form captured in semen is precisely that which is explained in the terms of the homogeneous parts: concoction working on elements. The work of organising the material elements uses the material elements to change them. The fundamental change is from externally heated to internally heated. This happens by being heated. When this heating occurs in such a way that it becomes internalised and able to spread that heat by virtue of itself, the heat has become capable of doing the work of form, of making not just homogeneous parts but nonhomogeneous parts.

In *Generation of Animals* II.1, Aristotle echoes his concern from *De Anima* II.4 that fire cannot explain the work of the nutritive soul: 'And just as we should not say that an axe or other instrument or organ was made by the fire alone, so neither shall we say that foot or hand were made by heat alone.'[53] The nonhomogeneous parts are not merely the heating of homogeneous parts. These parts follow after the heat that forms blood becomes internalised so that it orders from within. This ordering from within is the soul. As this ordering from within, soul does not become some hovering force, but an internally ordering force captured by the internal capacity to heat.

In his essay on the explanatory roles of teleology and necessity in the science of nature that comprises *Parts of Animals* I.1, Aristotle continues the argument that concludes *Meteorology* IV, arguing that there are two causes: necessity and the final end.[54] Nature is understood in two ways, as material and as the nature of substance, which includes the efficient and final cause. Aristotle denies the argument of those like Empedocles who think a substance can be explained by the chance occurrences in development and those like Democritus who say that the configuration of a thing is what makes it what it is, since a dead person is configured as a living person, but is not in fact a person.[55] Both positions lack a purpose, which is required for a sufficient account of a thing, and so lacking this account, they cannot sufficiently explain what a thing is.[56] In natural things, the purpose is to fulfil the nature of the natural substance. Form works on material to actualise the natural substance, which is the end, and the efficient cause coming from the father moves toward the material in the mother to form a new natural substance, which will draw in nutrients to actualise itself.

---

[53] GA 734b28–30.
[54] PA 642a1.
[55] PA 640b35–6. See Mariska Leunissen, *Explanation and Teleology in Aristotle's Science of Nature* for a monograph-length argument that Aristotle treats the final cause as explanatory and always presented in conjunction with either the formal or efficient cause. The emphasis on the approach to knowledge in the opening to *Parts of Animals* that leads to the conclusion that the final cause is the first cause seems to straightforwardly support the view that the final cause is required to fully understand but is not exercising some ontological force.
[56] PA 639b12–16.

While giving material a causal role, Aristotle argues that the material cause is not sufficient to explain the regularity of nature. Between his critique of Empedocles and his critique of Democritus, Aristotle writes:

> For it is not enough to say what are the stuffs out of which an animal is formed, to state, for instance, that it is made of fire or earth – if we were discussing a couch or the like, we should try to determine its form rather than its matter (e.g., bronze or wood), or if not, we should give the matter of the *whole*. For a couch is such and such a form embodied in this or that matter, or such and such a matter with this or that form; so that its shape and structure must be included in our description. For the formal nature is of greater importance than the material nature.[57]

Emphasising the need for form in giving an account of an animal or a couch, Aristotle includes the material. In conversation with Empedocles and Democritus, Aristotle accepts a role for material, just not the only role, hence he says that shape and structure must be included, having described form as embodied in matter. Formal nature, not as shape, but as what animates the material, better describes what a thing is than its material and the arrangement of the material does. This role of material in explaining what a thing is, coupled with the form, is echoed in the *Meteorology* IV.12 passage.

Hypothetical necessity is what must be the material case for the final cause to be achieved. In *Parts of Animals*, Aristotle lays out his approach by explaining that parts need to be explained in terms of how they serve the final cause. Parts do not just happen to accomplish their end because of the material of which they are constituted, they are constituted of that material in order to accomplish that end. Yet as Aristotle explains it, certain materials are required in order for that end to be achieved. The problem with the materialists is not that they recognise material's power, but that they think material alone can explain why natural substance is what it is. In Aristotle's example, the woodcarver is more correct to say that the tools are the cause of what is formed than the physiologist to say that the air and earth are the cause, because the tools lead to the end. But the tools alone do not explain, just as the air and earth do not explain. The woodcarver in naming the tools as the cause would also have to explain 'why he struck his blow in such a way as to effect this, and for the sake of what he did so'.[58]

In *Parts of Animals* I.1, Aristotle lays out the argument for his method for explaining the parts of animals by explaining that while material is necessary

---

[57] PA 640b22-9.
[58] PA 641a9-13.

for parts to achieve the end, the material (alone) does not explain why there are those parts. The parts have to be explained with reference to the way they contribute to the function of the animal. Semen is a liminal case in precisely this sense. The material capacities of semen are such to enable it to bring forth life, and yet the semen does not bring forth life just because it happens to have a certain degree of heat. It has the degree of heat, formed as it is by the father, for the sake of animating the menses. It comes from the father for this purpose, and yet, it does the work because it has the right kind of material.

Mariska Leunissen distinguishes between two ways of thinking of how material contributes to an end. Primary teleology involves the realisation of a potential for form that requires the proper materials to achieve its end, which Aristotle calls hypothetical necessity. Secondary teleology explains how the animal uses material to develop parts that enable it to better achieve its end. The materials are what are available to the animal, rather than what are required to achieve the end. In Leunissen's terms, secondary teleology contributes to 'luxury parts' that allow the animal to function better.[59]

Semen itself seems like a 'luxury part' that the animal forms by drawing on the material that it has on hand that allows it to better achieve its end – to produce another like it – which makes the relation of its material to its function seem like one of secondary teleology. When Aristotle describes nature as a good householder, he explains that like a good householder, nature assigns the best material to the best parts, and what is left over to parts that are less important (for Aristotle, the hierarchy to which he analogises these parts is that of the Athenian household: the free man, the slave and finally the animal).[60] Semen is formed out of useful residue – not quite waste, but certainly on the level of lesser material. Semen itself, the source of life, seems to come from the lesser nutriment. It is a part that is formed from whatever is available to the animal, and in this sense, seems like a lesser part than the parts that are more central to fulfilling its end, like the heart and the stomach. In secondary teleology, the material achieves ends, directed by the efficient cause, but does the work through the material. Semen seems to be just that kind of part that works through the material, directed by the father but having to become capable of working through the material alone to achieve its end.

---

[59] Leunissen, *Explanation and Teleology in Aristotle's Science of Nature*, pp. 17–21.

[60] Leunissen argues that Aristotle offers an account in PA 696b25–34 that runs against the view in the *Politics:* '[H]ere the "lower" animals benefit from the eating habits of "higher" animals. This indicates that there is no absolute subordination of the good of one living being to that of another', in *Explanation and Teleology in Aristotle's Science of Nature*, p. 45.

On the other hand, primary teleology would seem to be at work here as well. Semen has to be made of certain materials in order to be able to do what it does. Vital heat is not just the material available, it is the source of life. In order for semen to do its task, it must have vital heat. The πνεῦμα of the semen is the heated air. Semen is both formed out of a hypothetical necessity and from the materials that are left that can do the job to make reproduction more successful. Both a strong necessity and an almost incidental one seem to be at work at once. The first from the perspective of the semen's work in the offspring, and the second from the perspective of the semen's formation in the father.

How Semen is a Causal Part

Concoction differentiates the male contribution from the female contribution in a material process that produces the distinction between form and matter. If heat distinguishes the male and female contribution, and the male contributes form and the female material, it would appear that a material force is the cause of the distinction between what is material and what is formal. While Aristotle disavows the view that heat can organise the material into nonhomogeneous parts, the vital heat of semen is what enables it to bring soul into the material.[61] The problem is, as Cooper explains in his advocacy for the restricted material position, heat can explain why wood burns, but not how blood becomes a dog.[62] The higher animals need a force to move them, and this force is soul, but even this force operates through the heat of the soul, θερμότητος ψυχικῆς, since 'heat is a motive force'.[63] Semen requires certain material to do its work, and this material enables it do its work. Unlike other material parts, the function of this material part is to do the work of soul. Semen is the case where the material work seems to slide into formal work and the formal work into the material.

Aristotle himself struggles with this interrelation, trying to resist the reduction of soul to heat while still explaining the work of form in generation in terms of heat. Against the materialists, he writes in *Parts of Animals:*

> For some writers assert that the soul is fire or some such force. This, however, is but a crude assertion; and it would perhaps be better to say that the soul is incorporate in some substance of a fiery character. The reason for this

---

[61] *Meteor.* 390b3–11.
[62] John M. Cooper, 'Aristotle on Natural Teleology', p. 215.
[63] *GA* 732a16–19.

being so is that of all substances there is none so suitable for ministering to the operations of the soul as that which is possessed of heat. For nutrition and the imparting motions are offices of the soul, and it is by heat that these are most readily effected. To say then that the soul is fire is much the same as to confound the auger or the saw with the carpenter or his craft, simply because the work is done when the two are near one another.[64]

Fire is most suitable to do the work of soul, and yet, soul is not equal to fire. The analogy to craft that produces distance between the soul and its workings as it is taken up in *Generation of Animals* will be considered in greater length in Chapter Seven. It is clear in this passage that Aristotle both resists reducing soul to fire and describes its work through heat. The heat that is in semen is capable of a particular kind of movement, one that brings forth life. This movement originates in 'the movement set up in the male parent, who is in actuality what that out of which the offspring is made is in potentiality'.[65] Movement that originates in the father's particular form is now in the semen (τὰ σπέρματα). Through this movement, the semen works the blood up to the point where it has soul, animation, breath and eventually, the capacity to impart such a capacity into the offspring. While the heat that forms homogeneous parts seems like external heat following the account from *Parts of Animals* that blood's heat is external to it, the heat in semen belongs to it καθ' αὐτό. The source of this heat that each living thing has will be the source of life.[66] Aristotle juxtaposes what is external to what exists 'in the seminal fluid and the semen'. Against what is external, whatever exists in the semen, Aristotle explains, is either the soul, a part of soul, or contains soul.[67]

As Aristotle writes:

Everything that comes into being or is made must be made out of something, be made by the agency of something and must become something. Now that out of which it is made is the material; this some animals have in its first form within themselves, taking it from the female parent, as all those which are not born alive but produced as a grub or an egg; others receive it from the mother for a long time by sucking, as the young of all those which are not only externally but also internally viviparous. Such, then, is the material out of which things come into being, but we now are

---

[64] PA 652b8–15.
[65] GA 734b35–6.
[66] GA 736b34–737a3, see 766a17–20.
[67] GA 733b33–734a1.

inquiring not out of what the parts of an animal are made, but by what agency. Either it is something external which makes them, or else something existing in the seminal fluid and the semen; and this must either be soul or a part of soul, or something containing soul.

Now it would appear irrational to suppose that any of either the internal organs or the other parts is made by something external, since one thing cannot set up a motion in another without touching it, nor can a thing be affected in any way by anything that does not set up a motion in it. Something then of the sort we require exists in the embryo itself, being either a part of it or separate from it. To suppose that it should be something else separate from it is irrational. For after the animal has been produced does this something perish or does it remain in it? But nothing of the kind appears to be in it, nothing which is not a part of the whole plant or animal. Yet, on the other hand, it is absurd to say that it perishes after making either all the parts or only some of them. If it makes some of the parts and then perishes, what is to make the rest of them? Suppose this something makes the heart and then perishes, and the heart makes another organ, by the same argument either all the parts must perish or all must remain. Therefore it is preserved. Therefore it is a part of the embryo itself which exists in the semen from the beginning; and if indeed there is no part of the soul which does not exist in some part of the body, it would also be a part containing soul in it from the beginning . . .'[68]

In this passage at the beginning of *Generation of Animals* II, Aristotle posits the turn from that out of which a natural substance is made and that by which it is made. Aristotle focuses on nutrition in describing that out of which a natural substance is formed. Nutrition makes the substance grow. Nutrition is that out of which the blood that becomes menses comes to constitute enmattered substance having life.[69] The source of nutrition continues to be material from the mother, while the semen does not do this kind of nutritive work, which is why its material can evaporate. Not being nutrition does not mean that the semen's material composition is irrelevant to its work, as commentators argue.

In considering how the semen does its work, Aristotle recognises that the parts of the body cannot be made by something external to it. What moves needs to have some kind of contact with another to move it and what is moved cannot be affected without something that causes the movement. This cause,

---

[68] *GA* 733b25–734a16.
[69] *De An.* 415a24–7.

Aristotle argues, is in the embryo either as a part or as separate. It does not make sense to think it is separate, since the separate thing would become a part of what it causes, and yet such a thing is not observable in the embryo. Neither does it make sense for the cause to perish as a cause of motion, which would suggest that whatever causes motion to form the next part could and would perish, yet the organs must remain for the body to thrive. Aristotle concludes that a part of the embryo exists in the semen from the beginning.

He claims not that the semen exists in the embryo, but that a part of the embryo exists in the semen. What Aristotle then says is that 'if indeed there is no part of the soul which does not exist in some part of the body, it would also be a part containing soul in it from the beginning'. If there is no part of the soul that does not exist in some part of the body, then a part containing soul must be included in the body from the beginning. All of the soul must exist in some part of the body. The part of the embryo that exists in the semen must be the soul, which means that semen is the source of what organises the body. The body is the tool that manifests the soul. The human body is that whereby we are human. The soul is the ordering principle that makes the human body what it is. Semen works as the source for this ordering power to be internalised in a way that transforms the menses into the embryo that will nourish itself by continuing to draw nutriment from the mother in order to become the adult animal.

This account of soul explains why Aristotle must reject the pansomatist or pangenesis and preformationsist views of the way semen works. These accounts fail to fully appreciate the difference between actuality and potentiality. They fail to understand soul as the ordering principle for the whole that brings forth life and think of semen instead as carrying a collection of parts. On the pansomatist accounts, the semen was formed from all the parts. This account neatly explains how resemblance is possible, but it is unable to explain how the parts are unified in the body that is the offspring. As Aristotle argues, semen does not contain all the parts because if it did, something prior to semen would be required to unify all the parts in semen, which would lead to a regress of the cause and the question of how it causes. If the cause is semen, semen cannot carry the parts.[70]

Earlier in *Generation of Animals* I.18, Aristotle considers whether the semen is drawn from the homogeneous parts or the heterogeneous parts. Arguments for resemblance are used to support the view that semen is drawn from the heterogeneous parts (the face and hands and so forth). Aristotle argues that resemblance is not evidence for pansomatism because resemblance can be found

---

[70] *GA* 734a35–734b3.

in parts which nothing can come from, like voice and hair and nails, and children can inherit characteristics from grandparents who are contributing no part.[71] The alternative is that the semen comes from homogeneous parts (the flesh and bone and sinew). While this seems more likely, if the semen comes from the elements, from the very basic constituents, then it could not do the work of explaining how it became a whole. It would be whatever brought it together, whatever unifies the elements, not the most basic part, that was the source of its generation:

> If again something creates this composition later, it would be this that would be the cause of the resemblance, not the coming of the semen from every part of the body.[72]

The argument for pangenesis or pansomatism is that it would explain resemblance both to the parts and to the parent. That argument fails precisely because it fails to explain resemblance, which exceeds what could possibly be contributed. It also fails to explain how these contributions from all the various parts are unified to form the semen. Pansomatism views the natural offspring as a collection of parts, and semen comes from each of these parts and gets its capacity to produce resemblance from them. On this view, semen still needs some other cause to unify and animate what it causes.

Preformationism like pansomatism focuses on the parts and not the whole, and supposes that the parts are potentially included in the semen and caused in succession. Aristotle argues against this view that the semen causes the parts in succession, where one part is formed from semen, which then forms the next part because then each part would be contained in the previous part potentially, the liver being potentially in the heart and so forth, a view that Aristotle calls 'strange and fictitious'.[73] On this view, there is no ordering principle: the inclusion of each part in the one that comes into being prior to it causes the part it includes in due course, as if the semen were an encased set of dolls. Aristotle views the chief work of the semen as unifying and ordering, which is another way of saying animating. It works by departing from that of which it is the principle and joining with that whose principle it becomes.

> As, then, in these automatic puppets the external force moves the parts in a certain sense (not by touching any part at the moment, but by having touched one previously), in like manner also that from which the semen

---

[71] GA 722a3–7.
[72] GA 722b1–3. See further argument against this view at GA 764a1–765a4.
[73] GA 734a34.

comes, or in other words that which made the semen, sets up the movement in the embryo and makes the parts of it by having first touched something though not continuing to touch it.[74]

The semen comes from the father, who forms semen by concocting blood to the point where it is capable of movement that, like the father, can move in another to make the other capable of further moving in another. This concocted blood that is semen moves in the not-completely-worked-up blood that is the maternal menses to bring it to life in such a way that transforms it into an embryo and to organise the body by forming heterogeneous parts. The motion that happens in the father produces a residue capable of producing the motion that brings forth life. This heat becomes internal to the semen unlike the blood which the semen previously had been whose heat is external. This same blood courses through our veins, male and female, made hot by the heart. In the effort to pass on the form of the father to the form of the offspring, the form has to be both from the father and its own cause of the new offspring, if form is to be causal of individual substances. Semen becomes its own capacity for movement, which it gets from the father, but internalises into itself, by contrast to blood, which gets heat from the animal but does not internalise it.

The material basis of semen's work seems clearer in the account of how semen concocts – bringing the sufficient degree of heat to master the matter – than in the account of how the father concocts semen – by moving it through heat to become capable of moving in itself and in another. The latter case emphasises the moving cause that forms and en-forms semen, making the form that remains external and prior to the form formed by the semen the primary cause. Semen is like a puppet because it departs from the natural substance that causes it, having internalised the capacity of the parent.

This account of how the semen moves is not the same as having parts caused in succession where each part is included in the previous one potentially. Instead, the image of the puppet points to how motion moves in the parts.

> Now the semen is of such a nature, and has in it such a principle of motion, that when the motion is ceasing each of the parts comes into being, and that as a part having life or soul. For there is no such thing as face or flesh without soul in it; it is only homonymously that they will be called face or flesh if the life has gone out of them, just as if they had been made of stone or wood.[75]

---

[74] GA 734b12–17.
[75] GA 734b22–6. Cf. *De Motu An.* 703a29–703b2.

Semen is the part that makes the substance what it is, the part that forms and animates. Semen forms that part which when separated makes everything cease to be what it was. Semen moves to internalise its movement in that which through this movement becomes the ensouled embryo. Having internalised its movement in the embryo, when semen ceases to move, the parts that have been formed have life within them. The movement of the semen is capable of moving the parts so that they become capable of moving themselves, having soul. Each part is formed through what existed previously, the semen produces motion from its source, the kind of automaton that departs from that which sets it in its course to the point where it becomes capable of moving itself and moving in such a way as to cause movement in the other.

The image of the puppet seems to make the materiality of the semen not itself causal, but only causal in light of the form that sets it to work. But the relationship between form and matter appears more reciprocal in this process. Aristotle joins the image of the puppeteer to the analogy of building, where the focus in each is on the motion that does the work. As Aristotle writes, 'In a way it is the innate motion that does this, as the act of building builds the house.' The act of building actualises the passive potential in the building materials to be built into a house. As the materials are being worked on, this motion of building is in a way responsible for the house. In another way, the father and the builder are responsible. In building, the architect has the form of the house in mind and the builder moves the tools to form the building. The motion of the semen, by contrast to both the architect and the builder, occurs in light of its material constitution, a constitution made possible by the father, but now captured in the material as the material's own movement. This material capacity enables the semen to be independent of the father, more like a living creature, as Aristotle describes it in *De Motu Animalium*.[76]

In the passage that follows Aristotle's comparison of the semen to the puppet and to the builder, Aristotle situates a unique principle of motion in the semen, a principle that is able to produce a part that is itself capable of producing motion, having soul. This motion cannot be fully explained through the heat. Aristotle writes:

> [J]ust as we should not say that an axe or other instrument or organ was made by the fire alone, so neither shall we say that foot or hand were made by heat alone. The same applies also to flesh, for this too has a function. While, then, we may allow that hardness and softness, stickiness and brittleness, and whatever other qualities are found in the parts that have life and soul,

[76] *De Motu Anim.* 703b25–6.

may be caused by mere heat and cold, yet, when we come to the principle in virtue of which flesh is flesh and bone is bone, that is no longer so; what makes them is the movement set up by the male parent, who is in actuality what that out of which the offspring is made is in potentiality. This is what we find in the products of art; heat and cold may make the iron soft and hard, but what makes a sword is the movement of the tools employed, this movement containing the principle of the art. For the art is the starting-point (ἀρχή) and form (εἶδος) of the product; only it exists in something else, whereas the movement of nature exists in the product itself, issuing from another nature which has the form in actuality.[77]

A certain degree of heat achieves the concoction that enables the semen through the movement that heat causes to produce that movement in the maternal material to the extent that parts that can move themselves are formed. This passage is one that supporters of the restricted material view invoke to question the view that fire alone can do that work. Aristotle argues that heat can produce certain qualities in parts that have life, but heat cannot be the principle by which particular parts are particular parts. Not the heat of the tools but their movement is what makes the sword, Aristotle explains through his analogy to τέχνη. The heat is the means whereby the movement has an effect on the formation of the sword. This movement is a directed and guided movement with a purpose to form a tool that will accomplish a certain function. Semen similarly uses heat to accomplish a certain movement. But the heat that forms the sword so that it can move is not constitutive of the sword in the way that the heat that causes the movement of semen is. The sword analogy suggests that there is some form outside of semen that guides its movement through the use of heat. Having employed the analogy to τέχνη, Aristotle contrasts the work of art, where the ἀρχή and εἶδος of the product is in the mind of the artisan, to nature, where 'the movement of nature exists in the product itself'. This movement originates from that which has the form in actuality and thus can form through its movement the form in potentiality. The puzzle is whether the process of natural generation is more akin to τέχνη or φύσις.

Aristotle describes what happens in natural generation as the same thing that happens in artifice: 'This is what we find in the products of art.' Like art, flesh has a function. Having a function, we can conclude that it was formed for that function and does not have that function incidentally because of the material that constitutes it. Art is a clear example of what is formed for a function; the analogy explains how the material alone cannot do the work, but some organising force must do it, which in art is the movement of the tools.

---

[77] GA 734b28–735a4.

Aristotle notes the disanalogy is that the movement of nature exists in the product, which comes to be in the entity from another that has it in actuality. Material alone cannot make something purposeful, yet the purposefulness is intimately connected to the material substance in which it emerges.

On the τέχνη analogy, the heat accompanies the work of the tool – the iron tool has to be hot to shape the sword – but it is the movement of the tool, not the heat, that is the source of the shaping of the sword. For the natural being, this shaping and organising is internal, which explains how heat itself is the source of motion, both as material and as form. Heat directs the movement and heat comes from out of the material constitution. Because the form in τέχνη is in the art in the mind of the artisan not in a substance of the same form, a tool must do its work. Because the form of a substance originates in a similar actual substance, that substance does the work but still must pass from the substance to the new substance, and this occurs through something like a tool. The actual natural substance works on a potential natural substance that is like it, which is what the actual substance of the father is doing on the menses through the movement of semen, which is possible by means of heat.

This indistinction between the formal work of heat and the material of heat, between semen as internalised movement and movement that comes from the father, is also at work in the way that Aristotle describes semen as a cause of the parts. When Aristotle explains the way that semen is a cause of parts as a potential for them, he points to how semen forms through its motions which potentially possess the parts: 'Plainly, then, while there is something which makes the parts, this does not exist as a definite object, nor does it exist in the semen at the first as a complete part.'[78] The cause of the parts is not a 'definite object', nor is it a part in the semen, but rather the motion the heat of the semen motivates, which is the motion that formed the semen that the semen takes over from the father. The semen creates the parts through its movements as an outgrowth of creating life, not by having particular parts within it:

> Accordingly it is not any part that is the cause of the soul's coming into being, but it is the first moving cause from outside. (For nothing generates itself, though when it has come into being it thenceforward increases itself.)[79]

Semen is the particular part from outside – the part that comes from the father but becomes internal having its heat in virtue of itself and not externally as the menses as blood would – that moves in the menses to cause life. This argument for how semen is a source of motion on the basis of heat and

---

[78] GA 734b18–19.
[79] GA 735a12–14.

only as such is the source of life in the embryo indicates that Aristotle rejects the preformationist view that the parts of the the father pre-exist in miniature in the semen and are inherited by the embryo.[80] Against the preformationist view, Aristotle describes semen as a residue with a particular power of motion.[81] As a residue from the father, semen's work is not in being that of which the embryo will be comprised, though it does its work due to the material that it is.[82]

In the context of explaining the fact that male and female do not unite in all species that have male and female because the active and the passive need not join, Aristotle writes:

> It is plain that it is not necessary that anything at all should come away from the male, and if anything does come away it does not follow that this gives rise to the embryo as being in the embryo, but only as that which imparts the motion and as the form, so the medical art cures the patient.[83]

Even though semen brings no material to become a part of the material constitution of the embryo, the material constitution of semen can be of significance in generation without the semen contributing material. Further, Aristotle continues in the book that follows to consider the material composition that enables semen to be the source of motion.[84] The semen brings the soul, which organises the material to achieve the function of living. Once the semen does its soul-bringing work, the embryo internalises the life-capacity that makes the material do the work to achieve its end, no longer distinguishing between the material and formal contributions. That semen does not serve as the material does not indicate that its material composition plays no role.

The role of heat in animating is further attested by Aristotle's discussion of that first organ which when formed permits the semen to stop working.

---

[80] L. Aryeh Kosman argues that Aristotle's account makes the male the source of fertilisation but not the source of form in contrast to the female as the source of stuff in 'Male and Female in Aristotle's *Generation of Animals*'. As Kosman writes, 'The view that animals get formed by a successive series of motions that take place in the mother is at the heart of Aristotle's epigenetic theory, and the claim that that series begins with the motion of the sperm is what connects the sperm to that epigenetic theory. To suppose that the sperm operates by containing the substantial form of the animal in question is exactly to forsake epigenesis in favour of a theory that is essentially preformationist, even if it is what we might call formal rather than material preformationism', p. 241.

[81] *GA* 724b29–725a1, 729b5–20.

[82] *GA* 737a6–7.

[83] *GA* 729b18–20; cf. 764b10–14.

[84] There is a discussion of τέχνη at the end of this chapter (*GA* I.22 and I.23), but note that this discussion is of those living things where the male does not emit semen, or in plants.

The part that becomes the first principle of movement is the heart.⁸⁵ The heart is the body's source of warmth. When the heart is formed, the external source of vital heat has been internalised.

> The heart then and the liver are essential constituents of every animal; the liver that it may effect concoction, the heart that it may lodge the central source of heat. For some part or other there must be which, like a hearth, shall hold the kindling fire; and this part must be well protected, seeing that it is, as it were, the citadel of the body.⁸⁶

The chief organs are those organs that foster or maintain heat. The heart is the hearth that holds the 'kindling fire', which is the fire that is the source of continued heat. Semen has done its work when the heart is formed because the heart fosters heat on its own for the sake of the whole body. Aristotle will ascribe the task of 'work[ing] upon and concoct[ing] the nutriment' to all parts of the body. But,

> the source of heat in all the other members depends on this, and the soul is, as it were, set aglow with fire in this part, which in sanguineous animals is the heart and in the bloodless order the analogous member. Hence, of necessity, life must be simultaneous with the maintenance of heat, and what we call death is its destruction.⁸⁷

As the semen carries the soul, so the heart by its heat sets 'aglow with fire' the soul, and only when this heat in the heart is extinguished does death ensue.

In *De Anima*, Aristotle argues that fire alone cannot be the source of nourishment and growth, but soul must be.

> To some, the cause of nourishment and growth seems simply to be the nature of fire, for it alone of the bodies or elements manifestly nourishes itself and grows. Thus one might take this to be what is acting both in plants and in animals. However, while it is in some way a co-cause, it is not the cause simply, but rather soul is. For the growth of fire is without limit, as long as there is fuel; but of all things composed by nature there is a limit and a λόγος of size and growth – and these things are characteristic of soul but not of fire, and of λόγος rather than of material.⁸⁸

---

⁸⁵ *GA* 735a22–5. See also *Juv.* 468b28–469a9, 474a28–474b3. This text is considered spurious, and my argument, while supported by it, does not depend on it.
⁸⁶ *PA* 670a23–6; cf. *De Motu An.* 703a12–15.
⁸⁷ *Juv.* 469b15–20.
⁸⁸ *De An.* 416a10–19.

If soul is the principle of actuality, the principle of the order of the body, then this principle must exist in something else that causes the new body to be ordered. This principle of order in the body rather than the body does the work of animating. This new potentiality is actualised through the existing actuality. In art, the cause is not another actual existing thing of the same form. In nature, this actuality is a force of organising, it is not free-floating, existing in a mind or elsewhere as in art.[89] When the heart stops producing heat, the soul is no longer working to organise the body as body. The heat itself both is and is not the cause of the life of the body. It is that whereby the soul works on the body and when it ceases the soul has ceased – there no longer is an essence to the body that made it what it was. The soul is 'the origin of motion (ἡ κίνησις αὐτή), it is the end, it is the essence of the whole living body (τῶν ἐμψύχων σωμάτων)'.[90] The soul is the actuality, the very being or essence and the source of motion of the living body. Soul is actualised by semen. Heat is that whereby the semen moves. The concerns with the soul needing to be otherwise than either the semen or the motion or the heat each seem to make of the soul some homunculus-like figure that exists as fully formed entity. But the soul neither exists separately from the body, nor works by inhering in a receptacle that it defines. The soul is the ordering principle that makes the body appear as body. Semen is that whereby that principle moves from the father to the menses in which it works. The material of the semen is the cause of it being able to do its work, but only because the material allows it to do the work whereby soul can move it. This view does not entail semen contributing material. It does not deny the work of soul. Neither does it make soul into something that can only be forceful if conceived as without material power.

Material Composition of Semen

Aristotle explains that semen, 'in and with which is emitted the principle of soul', has two kinds of principles: one that is not connected with matter when there is a connection to reason and therefore the divine and one that is inextricable from the matter.[91] The rational soul is not blended with the body, recall, not because the activity of reason is external to the body, but because reason has no bodily qualities. The rational soul is actualised in thinking, as the nutritive soul is actualised in growing, and the perceptive soul in perceiving.

---

[89] Contra Sophia M. Connell, who argues that male and female principles that pre-exist male and female organs and male and female contributions explain generation and sexual differentiation in Connell, *Aristotle on Female Animals*, pp. 279–80.
[90] *De An.* 415b11–12.
[91] *GA* 737a8–10.

The rational soul of the human is not thereby separable from the growing and perceiving human being. It does not come to be in some separate process whereby the rational animal that is human comes to be. In *Nicomachean Ethics*, Aristotle explains that the bodily concerns of the human being limit the continuous actualisation of the rational soul. Contemplation cannot continue without end because human beings must eat and sleep and reproduce.[92] Similarly, in *De Motu Animalium* Aristotle describes a continuum between the work of φαντασία and αἴσθησις, which are situated in corporal substance, and νοῦς, which depends upon and reaches beyond what is corporeal.[93]

Some commentators take Aristotle's reference to the divine in connection to reason to suggest that some kind of more divine material, like αἰθήρ, must be responsible for life. On this view, πνεῦμα or breath does the work of semen not because of its specific material capacities but because it is closer to the divine. This more divine material shows how semen works through its material capacities to cause life, explaining how the 'male contributes the movement', and justifies the association of male capacities with the divine and female capacities with the earthbound.[94] Indeed, the concoction in the male that enables it to be hot καθ' αὑτό seems akin to a divine power. Aristotle associates the soul with the divine by means of vital heat, which he says is in all air. This connection between the soul, the divine, heat and air explains how, paraphrasing Thales's line 'all things are full of gods', Aristotle can say 'all things are full of soul'.[95] Aristotle seems untroubled by the view that air contains the capacity for soul, a claim that recalls the Hippocratics' position that breath plays a role in reproduction. Aristotle draws this association between air and vital heat in a passage whose reference to the ease with which generation occurs in an enclosure of air and vital heat forming a frothy bubble recalls the birth of Aphrodite, who emerges from the sea foam that formed from Ouranos's genitals cast into the sea by Kronos.

This recourse to the divine follows from the view that semen as the source of form cannot be operating materially. Yet Aristotle goes a considerable distance to explain how semen does its work through the specific material that constitutes it. Aristotle describes semen as a combination of breath and water in *Generation of Animals* II.1. Observing how semen acts, Aristotle concludes that

---

[92] *EN* 1178a24–1178b8.
[93] *De Motu Animalium* 701a35–6, 701b17–25, 703b19–20. For a broad notion of the animal capacity of αἴσθησις (perception) that includes judgement and for φαντασία (imagination) that mirrors thinking in the human, see Charles Kahn, 'Aristotle on Thinking', and Martha Craven Nussbaum, 'The Role of *Phantasia* in Aristotle's Explanations of Action'.
[94] *GA* 730a27–9.
[95] *GA* 762a19–20.

it cannot be just water or earth. If it were just water, heat would not thicken it, though it does. This thickening could be explained by semen being earth, but then it would not liquefy as thoroughly as we observe semen to do.[96] What we observe of semen tells us something about what the material must be if it is to be characterised by heat. Aristotle considers other substances that seem elemental but are distinct from the four elements, like foam and oil. Foam becomes thicker and whiter with smaller and fewer bubbles the hotter it is, which is similar to oil, which thickens when it is mixed with breath, πνεῦμα.[97] In *Meteorology*, Aristotle says oil is solidified by neither heat nor cold, but thickened by both because 'it is full of air (ἀέρος)'.[98] Semen has some of these qualities of foam, so it must be water and breath – 'it has a quantity of hot πνεῦμα in it because of the internal heat'.[99] Aristotle associates breath with air in ascribing to both the work of thickening oil, and by explaining the composition of semen as water and breath in *Generation of Animals* and as water and air in *Meteorology*. In *Generation of Animals*, Aristotle explicitly calls πνεῦμα hot air (θερμὸς ἀήρ). Aristotle concludes, 'Semen, then, is a compound of breath and water, and the former is hot air; hence semen is liquid in its nature because it is made of water.'[100] In *Meteorology* IV.10, Aristotle describes both blood and semen as 'made up of earth and water and air'.[101] Strangely, semen seems to be comprised of all the elements *except* fire. The breath-nature of semen is that by which the heat comes to the menses and is produced by further concoction. In the passage where Aristotle distinguishes vital heat from fire, he explains that vital heat is not fire, but breath, the breath that is in the semen.[102]

The argument is not that breath does the work and water evaporates, as if breath is the more divine part that is immaterial and water is material. Rather, the semen does its work materially through this breath joined to water. The work happens because of material power: the semen is not contributing material, but works to bring forth life in light of what the material that constitutes it is capable

---

[96] GA 735b4–6. See also *Meteorology*, where Aristotle specifies that: 'Those bodies that are made up of both earth and water are solidified both by fire and by cold and in either case are thickened. Heat acts by drawing off the moisture, and as the moisture goes off in vapour the dry matter thickens and collects. Cold acts by driving out the heat, which is accompanied by the moisture as this goes off in vapour with it' (383a12–18). If semen were earth and water then heat would make it dry rather than moist.

[97] GA 735b10–15. Marwan Rashed argues that semen operates as an organ due to the work of the frothy bubble in 'A Latent Difficulty in Aristotle's Theory of Semen: The Homogeneous Nature of Semen and the Role of the Frothy Bubble', in *Aristotle's Generation of Animals: A Critical Guide*.

[98] *Meteor.* 383b24–5.

[99] GA 735b34–5.

[100] GA 736a1–2.

[101] *Meteor.* 389a19–20.

[102] GA 736b34–737a1.

of doing. If semen is the source of individual form, it is individuated from the form of the father by being form in this new substance. Semen having its heat internal to it differentiates it from the form of the father, whose heat is internal to him. Material does the work, but the work is for form. In this sense, both material and form in the internalised heat of the semen is the source of differentiation from other individual forms. As the discussion of necessity and teleology in the previous chapter establishes, semen must be made of this material in order for it to do the life-forming work. The material is significant in semen's effort to actualise the menses even if the male contribution is not material.

Consider again the passage in *Generation of Animals* II.3 where Aristotle discusses the heat whereby the soul comes through semen:

> Now it is true that the faculty of all kinds of soul seems to have a connexion with a matter different from and more divine than the so-called elements; but as one soul differs from another in honour and dishonour, so also the nature of the corresponding matter. All have in their semen that which causes it to be productive; I mean what is called ~~vital~~ heat [θερμόν]. This is not fire nor any such force, but it is the breath [πνεῦμα] included in the semen [σπέρματι] and the foam-like, and the natural principle in the breath [πνεύματι φύσις], being analogous to the element of the stars [τῷ τῶν ἄστρων στοιχείῳ]. Hence, whereas fire generates no animal and we do not find any living thing forming in either solids or liquids under the influence of fire, the heat of the sun [τοῦ ἡλίου θερμότης] and that of animals does generate them. Not only is this true of the heat that works through the semen [διὰ τοῦ σπέρματος], but whatever other residue of the animal nature there may be, this also has still a vital principle [ζωτικὴν ἀρχήν] in it. From such considerations it is clear that the heat [θερμότης] in animals neither is fire nor derives its origin from fire.[103]

While fire 'generates no animal ... the heat of the sun and that of animals' generates living things. The heat that generates animals is distinct from fire, which is without limit, because it is limited and directive when it emanates from the sun and from actual animals. This heat is the internalised heat that moves from individual form to become the individual form in another. This heat is not fire but breath, πνεῦμα, and as breath, is analogous to the heat in the stars, the element of αἰθήρ. This association of vital heat with breath as hot air might explain why Aristotle equates αἰθήρ in Empedocles with air, because Empedocles points to the vitality of air.[104] In *De Caelo*, Aristotle ascribes the source of warmth and light of the stars to the 'friction set up

---

[103] *GA* 736b29–737a6.
[104] Michael M. Shaw, 'Aither and the Four Roots in Empedocles'.

in the air [ἀέρος] by their motion'. Aristotle explains that since movement creates fire in wood and in stone and in iron, it can be expected to create fire in air, 'a substance which is closer to fire than these'.[105] Movement creates fire in various substances, and even more does it create movement in air, which is more like it. This account seems like an intensification of heat that brings forth life, much as the Hippocratic author describes:

> The hottest and strongest fire, which controls all things, ordering all things according to nature, imperceptible to sight or touch, wherein are soul, mind, thought, growth, motion, decrease, mutation, sleep, waking. This governs all things always, both here and there, and is never at rest.[106]

In a passage that echoes Heraclitus's aphorism that the thunderbolt steers all things, the Hippocratic posits a particularly hot and strong fire that is directive. Aristotle's directive heat is material heat, whose material capacities enable it to do its work. As internalised heat, it governs that within which it persists. This heat constitutes semen in some way not merely as fire does by making hot, but by making capable of being hot on its own accord.[107] As Aristotle says in the passage associating πνεῦμα with vital heat mentioned above, 'all πνεῦμα is soul-heat [θερμότητα ψυχικήν], so that in a way all things are full of soul'.[108] All air contains this soul heat. The question is not how some things are alive for Aristotle, it would seem, but rather, how some things are not. The answer is that their heat is not internalised, which is to say, they do not have breath.

Soul (ψύχη) is the animating breath (πνεῦμα), which is heat and moisture. Breath is air, which is hot and moist, the two elemental forces that are essential for life. Breath in the semen is capable of doing the 'kindling' that internalises the heat in the menses. *On Respiration* explains that respiration transforms air through bringing it into contact with an internal heat: 'When it enters the air is cold, but on issuing it is warm owing to its contact with the heat resident in this organ.'[109] Air becomes hot as breath in respiration. Movement in the semen stops when life has begun, just as the offspring becomes fully separated from

---

[105] *De Caelo* 289a20–4.
[106] *Regimen* 1.10. Bartoš maintains that the Hippocratic author equates the innate heat and the element of fire as the sources of living things in Bartoš, 'Aristotle and the Hippocratic *De Victu* on Innate Heat and the Kindled Soul', p. 299.
[107] Contra Connell, who maintains that the male contribution is only accidentally hot. Connell argues on the basis of the later acquisition of heat that Aristotle references in *GA* 747a18–19 that heat is supplementary. Connell uses evidence that the Hippocratics held similar views to support her claim that the heat is accidental in Connell, *Aristotle on Female Animals*, p. 271.
[108] *GA* 762a19–20. My translation, revised Platt and Peck. Cf. *On Breath* 483a30–5.
[109] *On Respiration* 480b4–5.

the mother in breath.[110] The movement from the semen becomes internalised in the heart, which has its own internal heat and becomes the internal source of the animation of the offspring at birth when the movement from the mother no longer is needed for the offspring to live. It is not simply that sufficient heat in the πνεῦμα of the semen makes it capable of being a moving cause, but that this particular kind of heat enables the semen to work to internalise the heat in the menses to the point that it becomes its own self-sustaining and directing living substance.[111]

Sources of Vital Heat: Stomach, Sun and Earth

In joining Thales's claim that everything is made of water to the work of soul through the transitive property of πνεῦμα, Aristotle anchors form's work in material. What is under dispute is how some material or elemental power like heat can do the work of bringing forth life. The scholarship on this area focuses on what it means to distinguish this heat from fire to raise it to the level of a formal cause.[112] In elemental generation, heat characterises both air and fire. The heat capable of animating is a natural or vital heat, which I have

---

[110] *On Breath* 483a13.
[111] Andrew Coles makes the argument on the basis of the process of pulsation that produces heat in the πνεῦμα of the semen that makes the semen the moving cause in Andrew Coles, 'Biomedical Models of Reproduction in the Fifth Century BC and Aristotle's *Generation of Animals*'. Tress argues that his view depends on pansomatism and dismisses the work of form in Aristotle in 'Aristotle against the Hippocratics on Sexual Generation'.
[112] Friedrich Solmsen notes a tension in Aristotle's biology between places where he makes fire capable of the kinds of concocting changes attributed to natural or internal heat and places where he rejects this equation in 'The Vital Heat, the Inborn Pneuma and the Aether', p. 121n21. In *Parts of Animals* Aristotle speaks of: fire as the tool the soul uses, concluding that all animals have an amount of this heat; breath as feeding the 'internal fire', where fire or cognates of it are repeated three times in association with the breath; and natural heat that concocts as coming from the soul, which is 'as it were, set aglow with fire', at *PA* 652b8-16, *Juv.* 473a4-9, and 469b10-19, respectively. But Aristotle rejects their identity in other places, as in the passage above.

Gad Freudenthal argues in *Aristotle's Theory of Material Substance* that vital heat carries the soul, the enforming capacity, and this heat is manifested in πνεῦμα. On this reading, the πνεῦμα in Aristotle is a unifying force that keeps a substance together rather than all the elements flying in separate directions, pp. 137-8. Vital heat then works on the blood in such a way as to transform it into πνεῦμα which remains in the blood, p. 125. Freudenthal argues against the view he attributes to Solmsen that πνεῦμα is the instrument whereby heat is carried through the blood. Such an interpretation makes πνεῦμα into a *deus ex machina* since it seems unrelated to any other part of Aristotle's physical theory, p. 108. Freudenthal's solution is to show that vital heat is not an instrument of the soul that forms, rather vital heat carries the enforming movement, which is to say that it is soul, at once an efficient cause and a formal cause, not merely an efficient cause or tool.

argued is internalised heat, neither merely fire nor divinely inspired heat. Fire is not limited on its own accord, but moves toward the outer limits. The natural heat that Aristotle describes would seem to be an internal principle of order or at least a means whereby such a principle orders because it is heat that is internalised and capable of becoming internal. Vital or life heat is from the sun and is found in 'whatever other residue of the animal nature there may be'.

Aristotle uses various language to describe this natural source of heat. He calls it καθ' αὑτό, οἰκεία or its own,[113] κατὰ φύσιν, according to nature, or other variations of natural including συνφύτος or connatural,[114] ἔμφυτος or innate,[115] φύσεως θερμότης or natural heat[116] (and φυσικὸν πῦρ or natural fire),[117] θερμότητος ψυχικῆς, the heat of the soul,[118] θερμότης ζωτική, living heat,[119] and θερμότης ἐν τοῖς ζῷοις or animal heat.[120] Hippocratic authors also refer to this kind of heat, which they consider the source of soul.[121] The Hippocratics twice use the phrase συμφύτῳ θερμῷ.[122] They describe sickness as due to air mixing with connate or innate heat and nourishment being consumed by this heat. Two passages in Aphorisms refer to ἔμφυτον θερμόν: one ascribes the greatest of this heat to growing natural things and the least to old men; the other explains that this heat is greater in the winter, which explains why more food is needed in that season.[123] The Hippocratics also refer to ἔμφυτον πῦρ, innate fire.[124] They speak of the soul as catching fire, ἐκπυρουμένη.[125] And they speak specifically of the capacity of the semen to kindle, ζωπυρέονται, the female's contribution in generation.[126]

---

[113] *PA* 649a1–2, *Meteor.* 379a24–6, 379b19–25, 380a2, *GA* 784b4–5.

[114] *GA* 784b7, *De son et vis.* 458a27, *Prob.* 860a34, 883a7, 909b15–16, 949b5, *Juv.* 469b7–8. G. R. T. Ross translates this phrase as connate heat, and J. I. Beare translates it as 'connatural heat'.

[115] *Meteor.* 355b10.

[116] *PA* 648a27, 650a15, *Juv.* 470a20–1.

[117] *Juv.* 474b13.

[118] *GA* 732a18–19, *Juv.* 478a16.

[119] *GA* 739b23, *Juv.* 473a11.

[120] *PA* 651a11–12.

[121] Here I follow Bartoš, 'Aristotle and the Hippocratic *De victu* or Innate Heat and the Kindled Soul'. Bartoš maintains that Aristotle was interested in the medical disputes of his day, many of which can be found in the writings of the Hippocratics.

[122] *Morb.* I.11, *Vict.* II.62.

[123] *Aphorisms* I.14, 1.15.

[124] *On the Art of Medicine.*

[125] *Vict.* I.25.8.

[126] *Vict.* I.9, I.29, *Aphorisms* V.63.

In the passage from *Generation of Animals* II.3, Aristotle distinguishes this heat whose source is in the sun and in living beings from fire.[127] This capacity for being internal is what makes it innate or natural, as belonging in some essential way to a living thing, as a heat that is of the soul or of living things. Aristotle is not inventing this association of heat with life, nor the association of heat with what organises and governs, nor the distinction between fire and some fundamental heat that is essential to life. He gives a metaphysical framework to the previous efforts to describe what heat seems to be doing in generation by describing its work in terms of form. Clarity on how heat does this work will show how form works in and through material in natural substance for Aristotle. How semen comes to have this heat, the sense in which it is external to the menses, and the way that it is at work in both the male and the female contribution will point to how natural generation is the process of like working on like to actualise a potential in what is acted upon.

Aristotle could say simply that the heat of the male contribution comes from the heat of the male body. But he says that the semen has been worked up by a certain kind of heat that enables the residue that becomes semen to be concocted to the point where it can concoct the female contribution. All living bodies have this heat. The heat in the parents is the source of heat for both semen and menses. It must be located in an actually existing natural substance in order to become the source of motion in the potentially existing natural substance. Aristotle assigns the source of this heat to the sun and to animals. Deciding that it comes from the sun more fundamentally than from animals, as some commentators do, would seem to make this heat something more than the elemental force, something necessarily external. Yet the sun is not the only primary source Aristotle lists in this passage: 'the heat of the sun and that of animals does generate them'.[128] Sun and earth are the source of vital heat for plants, and animals are the source of this heat for their offspring. A certain kind of movement that living bodies are capable of produces this vital heat and all animals need this heat in order to live. Vital heat in animals need not come from the sun, but from a similar existing animal capable of transmitting heat to another to form further life.

This internal heat governs the body once it is situated in the heart. Another place this internal heat is found is in the stomach of both male and female bodies because nutrition requires it. Nutrition occurs in plants and animals when heat concocts food into blood. In *Parts of Animals*, again following a view espoused by the Hippocratics, Aristotle calls the stomach in animals

---

[127] GA 736b35–737a5.
[128] GA 737a3.

'the internal substitute for the earth', since it is where food is concocted into blood, while the earth plays this role for plants:

> Now since everything that grows must take nourishment, and nutriment in all cases consists of moist and dry substances, and since it is by the force of heat that these are concocted and changed, it follows that all living things, animals and plants alike, must on this account, if on no other, have a natural source of heat; and this, like the working of the food, must belong to many parts ... But animals, with scarcely an exception, and conspicuously all such as are capable of locomotion, are provided with a stomachal sac, which is as it were an internal substitute for the earth. They must therefore have some instrument which shall correspond to the roots of plants, with which they may absorb their food from this sac, so that the proper end of the successive stages of concoction may be attained.[129]

As the internal substitute for the earth, the stomach is to the animal what the earth is for plants: its source of this life-bringing heat. Aristotle explains that nutrients are transformed into the body for the sake of growth through the same concoction which, when it has reached its highest level – the one where the heat masters the moisture – will form the bodily part that can bring this power to concoct into another – semen. The stomach produces a heat whose own motion concocts nutrients into blood. In that concoction, the heat is internal to the body, but not to the blood. Only the moisture that also has air, the semen, has this heat and its accompanied motion within it.

The stomach has the capacity to concoct in a similar way that the semen does, because the stomach affects the food that enters it just as semen affects the menses it encounters: like a rennet. Aristotle explains that the semen acts as a 'curdling principle', just as rennet does upon milk.[130] In a passage following the one associating heat with breath in *Generation of Animals* quoted in the section above, Aristotle explains that semen evaporates after forming the embryo just as the fig-juice that curdles milk changes the milk without becoming a part of it.[131] But it matters very much that the fig-juice is materially constituted as it is in order to do its work even though it does not contribute material to what it makes.

Rennet, Aristotle tells us at the end of *Parts of Animals* IV, is found in animals with multiple stomachs, ruminants.[132] Food historians speculate that

---

[129] PA 650a2–9, 23–7.
[130] GA 729a11–20.
[131] GA 737a10–16.
[132] PA 676a6.

rennet was discovered when ancient peoples stored milk in the stomachs of young sheep, goats or cows on long journeys – the movement, the heat, and the rennet in the stomach would work on the base of milk to harden it into cheese.[133] If milk needs motion, heat and rennet to become cheese, it seems that semen holds these together in one homogeneous substance to work on menses to form life.

In *De Anima* II.4, Aristotle connects the warmth of the stomach with the warmth of the ensouled, when he writes, 'All food must be capable of being digested, and what produces digestion is warmth; that is why everything that has soul in it possesses warmth.'[134] Similarly, in *Meteorology* IV.3, Aristotle describes the processing of food in terms of heating from the body.[135] As Studtmann argues, the semen owes its becoming semen to the same process whereby food is converted into blood.[136] This is the same process whereby we become angry and our blood remains fluid. This heat also appears to be present, as Studtmann notes, in females who can begin the generative process but not complete it as evidenced by windeggs.[137] Studtmann argues that degrees of this vital heat distinguish between nutritive soul, which finds its work in the stomach, and the perceptive and thinking soul.[138]

The stomach is a source of heat, but in a different way from fire, which only heats. Only heating, fire is without limit. By contrast to fire, the stomach and semen bring a heating principle into what is heated, though they differ by the ways that what is acted upon is actualised. The food is acted upon by the stomach to make the food a part of the body, contributing to growth. The

---

[133] Mélanie Salque et al., 'Earliest Evidence for Cheese Making in the Sixth Millennium in Northern Europe'.

[134] *De An.* 416b28-9. Mark Shiffman notes that Aristotle makes comments in *On Sleep and Waking* (456b5-6) that suggest he treated digestion in greater detail in a work that is no longer extant in his translation of Aristotle's *De An.*, 57n12.

[135] *Meteor.* 381b7-9. See also *Juv.* 474a25-474b24.

[136] Paul Studtmann, 'Living Capacities and Vital Heat in Aristotle', p. 370; see *PA* 651b11.

[137] *GA* 741a17-18, Studtmann, 'Living Capacities and Vital Heat in Aristotle', p. 373.

[138] Studtmann argues that degrees of vital heat explain the different organisational complexity of nutritive and perceptive organisms on the basis of his argument that vital heat in the biological works parallels the different types of soul in *De Anima*. The difference between male and female that less heat is required to maintain certain capacities than is required to generate them. Studtmann, 'Living Capacities and Vital Heat in Aristotle', p. 378. This view appears traceable to Freudenthal who argues that vital heat produces the *scala naturae* determined in Aristotle by the degrees of complexity of the soul in Freudenthal, *Aristotle's Theory of Material Substance*, p. 4. Freudenthal argues that there is then a difference of degree within vital heat, but not between vital heat and elemental heat, p. 110. See also Cristina Cerami, 'Function and Instrument: Toward a New Criterion of the Scale of Being in Aristotle's *Generation of Animals*'.

stomach works on food more like the contrary works on a contrary, transforming it into itself. The food becomes other than it was in order to become body, being potentially body. Food becomes blood and flesh from something which emanates heat in itself, but the blood is not hot in itself. This work of the stomach is that of the nutritive soul. Similarly, semen works on menses in a way that internalises the heat so that it becomes capable of emanating the heat itself. Both the stomach and the semen work on material by emanating heat. Semen acts in such a way that internalises this heat in that which is potentially capable of emanating heat to make it actually like the semen, emanating heat, full of soul. The menses becomes alive in the fetation, while the food becomes a part of the body, made hot by its dependence on the heating principle of the body. Food acts as a fuel for the heat of the stomach, since adding to the substance in growth keeps the stomach in its work.[139] Menses is not fuel for the semen, but potentially like semen because it is potentially alive. Actualised, the menses becomes the embryo. The heat in the stomach and the semen is similar; what is different is the way that the material it works on is actualised by its heat.

---

[139] This sense of being soul is why Aristotle can say that the internal fire, ἐντὸς πυρός, comes from food not breath in *On Respiration* 473a2–13. Breathing does not add to the heat in the way that food does, making growth possible.

6
Sex Differentiation, Inheritance and the Meaning of Form in Generation

'The female is, as it were, deformed'

The case for the import of semen's material composition for its capacity to animate could be easily affirmed were it not for some of the difficulties raised by Aristotle's discussion of the cause of sexual differentiation and hereditary resemblance in *Generation of Animals* IV. Those who oppose the view that material plays a substantial role in semen's work argue from *Generation of Animals* IV that though some passages point to the independent power of the female contribution to resist or to overcome the male contribution, the male contribution is a principle not a material. As a principle, the male contribution must act if not without regard for, at least without dependence on, the material through which and in which it acts. This chapter makes the case from the specifics of sexual differentiation and trait inheritance that the male contribution, while it is form, works as form on material through material capacities, specifically of heat. The female contribution can resist the work of the male contribution and can 'loosen' the male's ability to produce an offspring just like itself because the male is working through material in a process that results in resemblance proceeding toward parents and grandparents and it would seem to regress until the resemblance is not recognisable to ancestors but only as a human being, sharing the form.

Sexual differentiation seems to be the process whereby the female becomes the 'deformed' member of the species. The language that makes the female the deformed male or the monstrosity points to Aristotle's assumption that the male is the true measure of the human and the female an aberration. Yet the argument still entails a mutuality between form and material. For example, Aristotle explains that the female is like the eunuch who is mutilated only in a part but that partial mutilation affects the whole. His point is not so much that the female falls short of the end of the male, but rather that change to a part that can serve as a principle affects the parts around it and in some cases the whole body.¹ The change in the parts will change the offspring from male to female, where the body will cease to be able to work up the appropriate amount of heat, lacking the parts, and as a result the body will come to have more blood, as the female body does.²

Aristotle's language of mutilation and monstrosity follows from defining animals in relationship to an end that is shared by similar animals. Both particular animals and kinds of animals are measured by an end that animals of the kind and animals of similar kinds achieve. The female on this account is understood in terms of an end of which the male achieves but can fall short and the female approaches but falls short. While Aristotle discusses monstrosity in the process of inheritance in *Generation of Animals* IV, in *Generation of Animals* II, Aristotle writes, 'For the female is, as it were (ὥσπερ), a mutilated (πεπηρωμένον) male.'³ This passage follows Aristotle's consideration of 'the material of the semen', which he divides between what is inseparable from material and what is divine and associated with reason.⁴ And this claim is made in the context of considering the differences between the way that the male is a residue and the way that the female is. Aristotle announces that the embryo and the semen have soul potentially but not actually, and then he says that the female has potentially all the parts, but none actually, including the parts that differentiate the male from the female because the female can be the potential for that which has a different potential than she does, Aristotle explains, just as nonmutilated children can be born from mutilated parents.⁵ Aristotle explains how this is possible by explaining that the female is, as it were (ὥσπερ), a mutilated (πεπηρωμένον) male. He finishes this sentence, 'and the menstrual fluids are semen, only not pure; for there is only one thing they have not in them, the principle of soul'.

Aristotle uses the same language 'as it were (ὥσπερ)' that he uses to refer to animals like seals who seem mutilated by comparison to others of similar kinds – quadrupeds – because their back legs are more like fins than hind

---

¹ GA 766a21–9. See Robert Mayhew, *The Female in Aristotle's Biology*, pp. 59–60 and Jessica Gelber, 'Females in Aristotle's Embryology'.
² GA 766b19–25. For the view that the variation in parts enables the male to better heat the male residue making sexual difference a material difference, see Marguerite Deslauriers, 'Sex and Essence in Aristotle's *Metaphysics* and Biology', pp. 147–51.
³ GA 737a27–28. For consideration of monstrosities see GA 769b11–30, where monstrosity is defined as a kind of deformity.
⁴ GA 737a6–9.
⁵ GA 737a23–7.

feet. They are deformed if those kinds are considered the measure or goal, but the apparently mutilated parts perform another function equally well for the purpose of seals, which is to move between land and water.⁶ Similar cases can be made for moles because they have a place for eyes, but cannot see; lobsters because they use their claws to walk; dolphins and whales because they seem to be fish but have no gills; and bats, because, like seals, they live in between – on the ground and in the air, so they resemble both birds and quadrupeds.⁷ These animals could be considered deformities on the terms that associate them with a wider kind, but on their own terms, they use their parts in perfectly workable ways to achieve their specific ends. When Aristotle says that the female is, as it were, a deformed male, he is not calling the female deformed in virtue of herself, but in contrast to the capacities of the male, just as the seal seems deformed in contrast to the capacities of a land animal. But the seal can do things that the land animal cannot on the basis of those 'deformities'.⁸

Aristotle shows form itself to be capable of being mutilated, of falling short of the goal of bringing forth the sufficient heat to cause it to be capable of bringing forth this heat in another. In the case of inherited traits, if deformity is failing to achieve the end of what it is like, the form can be so loosened that it is unable to produce offspring that resemble it. Similarly, the male offspring, which has the capacity for form through its heat, is by virtue of that heat more often born defective than the female offspring.⁹ If deformity here is failing to achieve the end of what it is like, then form is, as it were, deformity in these instances in a similar way to how material is, as it were, deformity.

Aristotle describes the male and female contributions that he assigns the roles of form and matter as having the same source – residue – and the same

---

⁶ *HA* 498a32–498b3. Allan Gotthelf makes this case in 'Notes toward a Study of Substance and Essence in Aristotle's Parts of Animals II–IV', pp. 230–1, followed by Mayhew, *The Female in Aristotle's Biology*, p. 55. Cynthia A. Freeland rejects Gotthelf's case in 'Nourishing Speculation', pp. 172–4.
⁷ *HA* 491b26–34, 532b30–533a15; *PA* 684a32–684b1, 697a15–697b14.
⁸ Sean D. Kirkland argues that Aristotle's 'proto-phenomenological method' begins not so much with common beliefs, as ἔνδοξα is commonly translated, but with the appearances of things, with our immediate experience of the world, which then become the occasion and motivation for further investigation in 'Dialectic and Proto-Phenomenology in Aristotle's *Topics* and *Physics*'. This view of Aristotle's method explains how ὥσπερ, the 'as it were', signals that the appearance is not the full sense in which it is, which will be manifested by further investigation. The appearances of the female and the seal as apparently deformed are a starting point for this further philosophical investigation.
⁹ *GA* 775a4–6.

process – concoction – whereby they become what they are. The male contribution is more worked up than the female's. If mutilation means no longer capable of serving a certain end, the male is mutilated female in the sense that it has become hotter than the female so that it is no longer doing the work of material, but now of form. For Aristotle, the capacity to bring forth life prioritises the form over material, because form can complete what material cannot. But material can still do what it does on its own terms. Even more, it can contribute to various incompletions of form. Aristotle discusses material's contribution to form's incapacity on numerous occasions throughout the biological works, where insufficient heat or excessive heat leads to certain kinds of differentiation as when too much heat coming up from below disturbs intellect or sensation or where insufficient material limits the capacity of an animal, as when lack of sufficient blood prevents concoction, increases in heat need to be balanced out, or hooved animals of necessity have fewer horns.[10] Of course, the excess heat that leads to changes in the animal's body does not change the form of the animal but the whole substance. Still, if having more heat, which distinguishes semen from menses, changes the capacity to act *as form*, then it seems that heat is determining whether the animal will function as form or material in generation.

Form *per se* is not a kind of or 'as it were' deformity as material is, because the end of form is what measures both form and material. After explaining that the male is naturally hotter, and for this reason is more often born defective, because the excess heat make the male offspring move more often *in utero* which causes injury, Aristotle once again acknowledges the weaker and colder nature of the female, which he says makes the female as it were deformed – ὥσπερ ἀναπηρίαν.[11] Throughout Aristotle's account, specific instances of form are susceptible to this becoming otherwise on the framework that makes the capacity of form the goal, just as the maternal contribution is capable of producing likeness. It is the material working within form that makes form susceptible in this way.

Material's inability to activate itself makes it deformed when measured by the goal the male achieves – to bring forth life to form another. On the other hand, the semen's goal cannot be achieved independently from matter, not just from material menses but from the matter through which it works. Having been worked up to the point where it can bring life into another, it can no longer be actualised to become the embryo. Even the male falls short of the capacity to reproduce itself without another. No Greek thought even the male gods could

---

[10] PA 672b27-32, 668b11-15, 663a29-34, 663b32-664a3.
[11] GA 775a4-15.

reproduce without the material provided by the female, goddess or mortal. Aristotle establishes a teleology that both male and female animals and their respective contributions of semen and menses make possible. The semen as the male contribution does not just need to have the menses as the female contribution to achieve life in the embryo, the semen needs menses in order to achieve the goal of the semen.

How Sex Differentiation Explains the Work of Form and Matter

Sexual differentiation brings into view the ability of the male animal to be overcome and the female to overcome in a way that shows the co-implication of form and matter. At the beginning of *Generation of Animals* IV.1, Aristotle recalls his claim that the sexes are first principles of living things, animals and plants.[12] What follows is a consideration of how these principles are manifested in such a way that determines the sex of the offspring. Aristotle rejects the view that the differentiation comes through the seed of the male as it does for Anaxagoras, or occurs in the uterus based on the location of the offspring and the corresponding heat of it as it does for Empedocles, or occurs in the mother depending on whether the male or female seed prevails as it does for Democritus.[13] Many think it is due to heat, but Aristotle rejects this view if it only means the conditions of heat being in place, as if a hotter uterus could differentiate the sex of the offspring. After considering problems with thinking that heat alone or that parts alone are the cause of differentiation, Aristotle concedes that it is 'not altogether unreasonable' to suppose that heat and cold are the cause of male and female. The problem is that speaking in this way 'is to seek for the cause from too remote a starting-point; we must draw near the primary causes in so far it is possible for us'.[14] The heat and cold do the work, but this explanation is incomplete because it does not explain the primary cause – that which is working through the heat, we might say. Following this passage, Aristotle distinguishes the male and the female,

> by a certain capacity and incapacity. (For the male is that which can concoct and form and discharge a semen carrying with it the principle of form – by 'principle' I do not mean a material principle out of which comes into being

[12] *GA* 763b23–4.
[13] *GA* 763b28–764a22.
[14] *GA* 765a35–765b6.

an offspring resembling the parent, but I mean the first moving cause, whether it have power to act as such in the thing itself or in something else – but the female is that which receives semen, but cannot form it or discharge it.) And all concoction works by means of heat. Therefore the males of animals must needs be hotter than the females.[15]

The male has to be hot to concoct. The coldness of the female and its inability to concoct explain why she has more blood which enables her to contribute material to the offspring. Blood is hot, what has more blood is hotter. But the female contribution is not from excess blood, just as excess of the useful residue out of nutriment is not a result of a greater amount, but of a certain working up. The degree of working up distinguishes between the kinds of contributions.

Aristotle argues that these capacities correspond to particular parts in the male and female body – the testes in the male and the uterus in the female. The organ is the site of the function; a part of the body comes into being with the function the part achieves. The organ comes to be and increases its being from nutriment and each part is made out of the kind of material it draws on in order to grow. That makes it seem like the part grows from what is the same as it is, but in a sense, it forms out of its opposite, which Aristotle does not yet explain. If it comes to be from its opposite, it also decays into its opposite. Specifically, Aristotle says, 'that what is not under the sway of that which made it must change into its opposite'.[16] What generates comes to be, in a sense, from its opposite. The organ, for example, comes to be from nutriment and grows through nutriment. If it is under the sway of what made it, it remains the same. If it is not under the sway of what made it, it changes into its opposite. The context of this discussion is explaining the difference between male and female. Nutriment is what makes the organ. If the nutriment does not continue to grow the organ, it ceases to be the organ. As Aristotle explains, when a thing perishes, it becomes the opposite of what it was. Perishing is becoming the opposite. Aristotle concludes that what is 'not under the sway of that which made it' must change into its opposite. Aristotle has just associated becoming opposite with perishing. But he now describes this changing into opposite in terms of a thing not being under the sway of that which made it. In this context, the nutrition was the cause, that from which it came. The thing ceases to be under the sway of the nutrition

---

[15] GA 765b9–17.
[16] GA 766a14–15.

when it ceases to eat, just as the part disintegrates when the part ceases to take in nutriment, which is what made it.

With all this in mind, Aristotle writes:

> For when the principle (ἡ ἀρχὴ) does not bear sway (μὴ κρατῇ) and cannot concoct (μηδὲ δύναται πέψαι) the nourishment through lack of heat (ἔνδειαν θερμότητος) nor bring it into its proper form (τὸ ἴδιον εἶδος), but is defeated (ἡττηθῇ) in this respect, then must ~~the material~~ change into its opposite. Now the female is opposite to the male, and that in so far as the one is female and the other male. And since it differs in its faculty, its organ also is different, so that the embryo changes into this state. And as one part of first-rate importance changes, the whole system of the animal differs greatly in form along with it. This may be seen in the case of eunuchs, who, though mutilated in one part alone, depart so much from their original appearance (τῆς ἀρχαίας μορφῆς) and approximate closely to the female form (τοῦ τὴν ἰδέαν). The reason of this is that some of the parts are principles (ἔνια τῶν μορίων ἀρχαί εἰσιν), and when a principle is moved many of the parts that go along with it must change with it.[17]

Before we interpret this crucial passage for our understanding of the chapter, we should understand the conclusions Aristotle draws from it. In the following paragraph, he establishes 1) that male is a source (ἀρχή) and a cause (αἴτιον); 2) male is defined by having a capacity; 3) female by having an incapacity; 4) the capacity is the ability to concoct the nutriment to its ultimate stage, which is blood; 5) the cause of this capacity is in the source (ἐν τῇ ἀρχῇ), which is the part that contains the source of natural heat. If all these things are true, there must be a heart that is formed out of this heat. We know from other passages that the heart becomes the internalised source of heat in the body, formed out of the blood. When it is formed, the embryo is alive. The heat is the cause of the male and the female, and it resides in the part of the body that produces the heat.[18]

This argument for how the male does the work to produce the internal source of heat allows Aristotle to return to the language of what prevails and fails to prevail. The semen (σπέρμα) is the ultimate residue of the nutriment, which goes to all the parts and so makes life and resemblance to parents possible. Aristotle speaks specifically to the semen (σπέρμα) of the male which 'differs', implying that the previously discussed semen was of the male and female. This

---

[17] GA 766a17–29.
[18] GA 766a30–766b3, 735a22–5, 741b15–21.

male semen has a source within itself whereby it can begin the movements in the embryo (ἀρχὴν ἐν ἑαυτῷ τοιαύτην οἵαν κινεῖν). This source also concocts the ultimate nourishment of the seed, but the female seed contains only material and not this additional capacity.[19] What follows is the crucial sentence:

> If, then, the male element [referencing the male semen described as a principle of movement in the first clause of the previous sentence] prevails (κρατῆσαν) it draws (ἄγει) the female element into itself, but if it is prevailed over it changes into the opposite or is destroyed. But the female is opposite to the male, and is female because of its inability to concoct and of the coldness of the sanguineous nutriment.[20]

This account of what follows can now inform the interpretation of the passage above about the first principle not 'bearing sway' and something changing into its opposite. In that passage, Aristotle speaks of a source or principle, which we now see connected to a particular part that is the source of this function of movement.[21] When the male element does not master the nourishment, it – the male element – changes over into its opposite – the female element – because it is no longer acting as the male element when it is unable to bring forth the capacity to bring forth life in another. The male element that is working on the nutriment qua male, that is, through its power to draw in nutriment that is like it insofar as semen too develops from nutriment to produce another with this same capacity, produces a female when it cannot fully internalise the nutriment. It internalises the nutriment enough to make it alive, but not enough to make it capable of producing this heat in another. In this sense, the forming of the embryo and sex differentiation happen at the same time. The potential for the material to be male is converted into the potential for it to be female.

The potential for the material nutriment to be male is actualised by the potential in the semen. This actualising that is the work of the male semen occurs in the female material. When the embryo becomes female it is due to a certain degree of failure of the male semen.[22] The actuality which would have been a

---

[19] GA 766b4–15.

[20] GA 766b15–18.

[21] Sophia M. Connell argues that the references to male in this passage are to a principle because form is a principle and not a part or embodied capacity in Connell, *Aristotle on Female Animals*, pp. 306–8.

[22] Bos argues that sexual differentiation depends on quality of πνεῦμα in the semen, which Bos wants to associate with divine matter, but the move to πνεῦμα shows how the account requires material explanations in *Aristotle on God's Life-Generating Power*, p. 68.

living being capable of forming a living being in another – a male – becomes its opposite and in this sense the male element becomes its opposite when it is not doing the work of animating a living being capable of animating another.

In working on the female material and actualising its potential, while still being prevailed over by capacities in the material, the semen forms a female embryo – the being with an internal source of heat itself in the form of the heart, but not an internal source of heat capable of producing another with an internal source of heat. What is transformed to the opposite is the capacity to be the active potential, the capacity of the male semen. Having been prevailed over it can no longer do the work of producing another with the capacity to form another.

While earlier Aristotle has said that heat is too distant a cause to explain sexual differentiation, the central importance of the proper measure of heat and moisture returns in the next chapter, where Aristotle points to the way that heat and moisture in the environment influence sexual differentiation.[23] The young and the old produce more females because the heat in the young is still developing and in the old failing. Moister bodies produce more females because the moist overtakes the hot. Males with more liquid semen produce more females because their semen is not as concocted. As Aristotle explains, 'all these characteristics come of deficiency in natural heat'.[24] Hotter circumstances produce more males as do circumstances with less moisture because in these circumstances, it is easier to concoct. The material circumstances of the presence of heat and of the amount of moisture affect the sex of the offspring. As Aristotle explains, generation depends on the proper mixture of the hot and the moist. The analogy Aristotle uses is to cooking, where both too much heat and too little makes the process a failure.[25]

This same dynamic is in play in the account of resemblance in *Generation of Animals* IV.3. Aristotle describes the course of resemblance by noting a way that the offspring departs from type. The origin of resemblance occurs in generation when the menses is concocted. If it is properly concocted, the movement from the male through the semen (Aristotle says it makes no difference whether we are speaking of the semen or the movement since the formula of the movement (λόγος τῆς κινήσεως) is the same in the movement and the semen) makes the embryo resemble the male. Aristotle writes: 'Thus if this movement prevail (κρατοῦσα), it will make the embryo male and not

---

[23] *GA* 766b33–767a12.
[24] *GA* 766b27–33.
[25] *GA* 767a20–7. Connell argues that the language of cooking suggests that both male and female parts are doing equal work and the differences are based on different ratios, or συμμετρία in Connell, *Aristotle on the Female Animals*, pp. 271–5.

female, like the father and not like the mother; if it prevail not, the embryo is deficient in that faculty in which it has not prevailed.'[26] The same elision of what is worked on is in the second phrase as in the passages in *Generation of Animals* IV.1 above. In this chapter, Aristotle reminds the reader that things change into their opposite,

> Now since everything changes not into anything haphazard but into its opposite, therefore also that which is not prevailed over in generation must change and become the opposite, in respect of that particular force in which the generative and moving element (γεννῶν καὶ κινοῦν) has not prevailed. If then it has not prevailed in so far as it is male, the offspring becomes female.[27]

A certain movement must prevail or master and when it does not, the embryo becomes the opposite of what failed to prevail. What is not prevailed over – the material not being fully mastered – makes the embryo become the opposite of the generative and moving element.

The same process can be explained in two ways: the male fails to prevail or master and the female fails to be prevailed over or mastered. In both cases, the failure leads to a becoming opposite. What is in question is to what the movement is opposite. In the first case, 'that which is not prevailed over in generation must become the opposite', seems clearly to point to the female nutriment not being prevailed over. It becomes the opposite, not of female, but, to finish the sentence, 'in respect of that particular force in which the generative and moving element has not prevailed'. What is not prevailed over becomes the opposite, not in respect to what it was, but in respect of what the moving principle was trying to accomplish. 'If then it has not prevailed in so far as it is male, the offspring becomes female.' This sentence follows as a continuation of the earlier discussion, not an introduction of a new possibility: the female matter was not mastered, which is another way of saying that the male principle did not prevail. It is not that sometimes the female fails to be prevailed over and sometimes the male fails to prevail, but that in the same case the female fails to be prevailed over and the male fails to prevail. In that case when the male both is mastered and fails to master, the male principle is no longer acting as the male principle and so is transformed into its opposite. In neither case does an already formed male embryo need to become female.[28]

---

[26] *GA* 767b20–4.
[27] *GA* 768a1–6.
[28] Cf. Connell, *Aristotle on Female Animals*, pp. 306–10.

The change is from becoming the kind of living thing that can produce another living thing capable of producing a living thing to becoming the kind of living thing that can generate another living thing in itself but not bring life into it. Because male is defined as that which can generate in another, the male succeeds when the embryo is formed, but falls short and fails to master the embryo when the male is unable to concoct to the point where the embryo will be able to concoct in another. The male's failure is the flipside of the female contribution's activity that keeps the male from prevailing.

How Inherited Traits Explain the Role of Form

Aristotle inherits from the Hippocratics the tasks of explaining how generation is sexual, requiring both the male and the female, where he distinguishes the contributions in terms of form and matter, how it produces animals of two sexes within the same species, and how that distinction still allows male and female parents to contribute inherited traits. The Hippocratic authors were concerned with the inherited traits question, which is why they saw generation as a mixing of seeds that had come from all the parts of the body. Aristotle finds that view implausible because it does not explain how the contributions from all the parts come to be unified in the contribution of the parent, a role he ascribes to form, and hence to semen. Another possibility was that what was contributed was a miniature being with all the traits of the parent – a homunculus – which Aristotle also rejects because that would still need to be unified with the contribution of the other parent. Aristotle argues that a forming capacity comes from the father and a capacity to be formed comes from the mother. The father's contribution actualises the mother's contribution. When the father's contribution fails to fully actualise the mother's contribution, the offspring becomes of the same capacity as the mother, having a capacity to live but only to contribute what can be formed to the offspring. The movements that produce inherited traits occur in this same work wherein in generation the life is formed and its capacity to form life in another is determined. These are not distinct stages of movements, but these three processes of producing life, producing the capacity to generate life in another and producing resemblances occur if not all at once, in tandem. Otherwise, the male and female contributions would be remaining distinct after generation has occurred and this separation would separate the form and matter in natural substance after the process of generation, which would not make sense for Aristotle, whose account focuses on how the male and female contributions are unified in the fetation.

Resemblance poses a problem for understanding the role of form and material as associated with male and female in reproduction. Children can look like their mothers. If they look like their mothers, then mothers must have some role in inherited characteristics. If mothers have some role in inherited characteristics and the mother's contribution is material, then the inherited characteristics are material. If the inherited characteristics are material, then the father, contributing form, would not be capable of affecting resemblance, which he clearly does. One way to solve the problem is to distinguish between the role of individual form and the role of shared essence, or between the role of the individual and the role of form in order to allow for the role of the mother in resemblance. In *Generation of Animals*, Aristotle is trying to explain both how an individual can be what it is as an individual and how it can share the form of the parents and other members of the species. Efforts to resolve this problem have led to some original interpretations, for example, that all possible characteristics and their variations are included in the male contribution, because it is the form, while the female contribution occurs only when the male contributions fail to be actualised, or that the male is responsible for actualising the potentials in the female.[29] The solutions that find a place for the female must either argue that the mother contributes to form,[30] or argue that male and female are material differences.[31] If it is correct that the form should be viewed as individual and causal of substance, and the definition is of the many similar forms,[32] the mother cannot be the cause of form because then the form would be multiple in the substance. These proposed solutions point to how the question of whether form is individual or species-form and

---

[29] John M. Cooper argues on the basis of A. L. Peck's translation of *GA* 768a11–14 that includes the phrase 'of the female' to modify the potential movements in the semen, that no δύναμις, κίνησις or mastery comes from the female, which suggests that all potential characteristics that either the male or female could contribute must be included in the male in 'Metaphysics in Aristotle's Embryology', p. 30. For careful rebuttals of Cooper's view, see Heidi Northwood, 'Disobedient Matter: The Female Contribution in Aristotle's Embryology', and Connell, *Aristotle on Female Animals*, pp. 303–24. See also George Boys-Stones, 'Physiognomy and Ancient Psychological Theory', pp. 47–55, C. D. C. Reeve, *Substantial Knowledge*, p. 55 and Mary Louise Gill, *Aristotle on Substance*, p. 125n29.

[30] D. M. Balme argues that Aristotle posits individual forms that both the male and female potentially contribute which contain incidental features of parts in 'Άνθρωπος ἄνθρωπον γεννᾷ: Human is Generated in Human', pp. 27–8. See also Balme, 'Aristotle's Biology Was Not Essentialist', Devin Henry, 'Understanding Aristotle's Reproductive Hylomorphism' and Peck, *GA* xiii.

[31] Deslauriers, 'Sex and Essence in Aristotle's *Metaphysics* and Biology', pp. 138–67.

[32] Charlotte Witt, 'Aristotle's Essentialism Revisited'. For further support for individual form in generation, see Henry, 'Aristotle on the Mechanism of Inheritance', Montgomery Furth, 'Transtemporal Stability in Aristotelian Substances' and Anthony Preus, 'Science and Philosophy of Aristotelian Generation of Animals'.

whether formal and material contributions can produce effects that do not correspond to their role come to the fore in Aristotle's treatment of resemblance.

Arguments that explain the work of resemblance in terms of movements in the male and female and thereby defend species form in Aristotle's account of resemblance emphasise Aristotle's use of a craft analogy to explain the work of semen as a tool of form. Gelber argues that the movements, located in both the male and female contributions, are tools not of form, but of the generative parents, which allows the mother to have influence over the resemblance of the child.[33] Connell similarly argues for a distinction between the forming of the individual and the work of form as such. Her distinction is between the work of μορφή, which shapes the individual, and εἶδος, which is the cause in a more technical sense, making μορφή a tool of εἶδος.[34]

---

[33] Jessica Gelber, 'Form and Inheritance in Aristotle's Embryology', and 'Females in Aristotle's Embryology', pp. 182-6. Gelber separates the work of the embodied or material form from the principle of form and the movements of the generators – the male and female parents – from the movement of species form. The movements are nonaccidental with regard to the generators and accidental with regard to form. On this view, Aristotle is not discussing the work of form in *Generation of Animals* IV.3 since he never uses the word εἶδος in the chapter. See Gelber, 'Form and Inheritance in Aristotle's Embryology', pp. 196-7 and Witt, 'Form, Reproduction, and Inherited Characteristics in Aristotle's *Generation of Animals*', p. 49. Gelber argues that some features in individual substances are not part of the form, which is universal and so not subject to individual particularities. Contra Connell, who argues that if the inherited traits are all material, they would all need to come from the mother in *Aristotle on Female Animals*, pp. 300-2. Sarah Borden Sharkey agrees that some characteristics of substance are not formal, such as sexual characteristics, which she describes as part of the material conditions of the substance in *An Aristotelian Feminism*, p. 42.

[34] The purpose of this argument is to maintain the transcendence of εἶδος as the species form from the actually existing composite natural substance in Connell, *Aristotle on Female Animals*, p. 319. Yet Aristotle equates μορφή and εἶδος in *Physics* II.1 when he argues that form is the definitive principle of a natural substance. 'Thus on the second account of nature, it would be the shape or form (ἡ μορφὴ καὶ τὸ εἶδος) (not separable except in statement) of things which have in themselves a principle of motion' (*Physics* 193b3-4). What is separable in this passage, only according to the logos, κατὰ τὸν λόγον, is not the μορφή from the εἶδος which are synonymous, but the form from the matter. Support for this reading is in the next sentence, where Aristotle says that the combination of the two is not nature, but by nature. Human beings are a composite of form and material, and this composite is by nature, while the form is the nature of it. In conclusion to this argument, Aristotle twice uses μορφή to define nature (*Physics* 193b11-12, 18-19). It seems reasonable to understand μορφή as the principle that is doing the shaping and εἶδος as the form if we are able to see them as two ways of thinking of the same cause. The first is not a tool of the second, but the active work of that which is manifested when it is actualised. This conception of μορφή helps us understand semen as an active shaping that is the work of form or εἶδος without concluding that it is otherwise than form.

I have argued that the form of the father must work through the material of the semen in order to maintain Aristotle's position that natural forms are always actualised in substance, rather than floating properties that come to inhere in unformed stuff. On this account, the form that comes from the father is not itself the form of the offspring – a new form is at work in the offspring, which is distinct from the father by being the cause of a separate substance. The soul is 'relative to a body', and its activities are in a body. Gelber recognises that the movements of semen are material in order to permit both the semen's and the menses' movements to cause resemblance, but in order to do so suggests that semen is not soul but a tool of soul whose movements are somehow other than soul.

One way to think of semen as the tool of soul is to see it as that through which and in which soul is working, and in that sense semen is the manifestation of the activity of soul, bringing life into the menses to form the embryo. But if the semen is the tool in a way that is not itself form, it does not explain how form comes into the menses. If soul must be kept distinct from material semen, material semen must be a tool that can do the work of soul without being soul. The question remains of how soul passes to material semen in order to make semen capable of doing its forming work. If form is somehow carried into semen then it seems that semen does hold the power of soul. If form is acting in semen without semen itself being soul, how it does the work of soul in material remains unclear. Form either works as a transcendent force not needing to come from an actual existing natural substance – it is species' form and therefore not causal of individual substance – or it is individual form that is immaterial and must somehow jump a gap between the actually existing substance and the new substance without a material exchange. On this account, it would seem that menses is actualised by desiring the form of the father as Aristotle describes at the end of *Physics* I.9. If this is the case, no material would need to pass from the father to the menses.

Having established that the embryo changes from male to female when the male insufficiently masters the female contribution, Aristotle follows a similar structure to explain how resemblance can vary from male to female parent, grandparents, to generic resemblance of the form. He explains: 'Some of the movements exist actually, others potentially; actually, those of the father and the general type, as man and animal; potentially, those of the female and the remoter ancestors.'[35] In this passage, Aristotle distinguishes between the

---

[35] GA 768a12–14. Ἔνεισι δ' αἱ μὲν ἐνεργείᾳ τῶν κινήσεων, αἱ δὲ δυνάμει, ἐνεργείᾳ μὲν αἱ τοῦ γεννῶντος καὶ τῶν καθόλου, οἷον ἀνθρώπου καὶ ζῴου, δυνάμει δὲ αἱ τοῦ θήλεος καὶ τῶν προγόνων.

movements of the father and the general type that are actual and the movements of the female and the remoter ancestors that are potential. In the course of making this distinction, Aristotle reminds us that what loses its own nature turns into its opposite. The movements that form the embryo can relapse, λύονται, literally loosen, and in relapsing or loosening they do not change into their opposite, but into what is close, the male parent into his father, and if a little more of a relapse then into his grandfather. Similar relapses occur on the side of the mother for the female inheritances.

In a passage that suggests that the change into opposites with respect to sexual differentiation and resemblance occurs at the same time, Aristotle explains that the male movement (κίνησις) can prevail without the specific traits of the father prevailing.[36] When the male movement prevails, but the father's movement relapses, the offspring resembles the opposite sex parent. When further relapse occurs, the male movement can prevail but the movement from the individual father relapses into his father so the child resembles the grandfather. When the male movement is prevailed over, the child is female and increasing degrees of relapse result in resemblance of grandmother of the parent. Aristotle notes that these movements and relapses can occur at the level of individual parts. The movements can be so confused that a clear relapse to the next parent or remoter ancestor cannot be traced, and the child seems only to resemble the human being. 'The reason for this is that this accompanies all the individual characteristics; man is universal, while Socrates, the father, and the mother, whoever she may be, are individuals.'[37] Aristotle makes the case that the movement that is most actual without any relapse resembles the father, so it would seem that the individual form that is actualised without any modifications comes from the father, even though, being the cause of this new substance, it is different form from the father's form. Movement away from the individual form of the father occurs from a kind of shortcoming in the father's form being fully causal, which causes the resemblance to be only what is shared among those who have the same species form.[38] Witt argues against the view that this passage indicates that species form is causal of substance even if it is only after a series of relapses from individual form. Rather, Aristotle means by universal movements that whereby the offspring becomes

---

[36] *GA* 768a29–31.
[37] *GA* 768b12–14.
[38] Balme argues from this passage that the universal species form is not the primary cause of reproduction, the individual form is, a view that leads him to deduce that individual forms are unique and themselves the cause of the distinctions between individuals in Balme, 'Aristotle's Biology Was Not Essentialist'. Cf. Gelber, 'Form and Inheritance in Aristotle's Embryology', p. 194n28.

one of a species. When the individual movements of the parents have been so loosened, the resemblance is restricted to the kind and does not include resemblance to the parents.[39]

Aristotle explains how the movements of the father can be affected by the movements of the mother: the actual movements of the father are loosened, so the resemblances are not quite as actual as the parent, but are of the parent's parent. When, in the process of generation, the female contribution is not fully actualised, and the offspring becomes female, the potential movements of the female have some sway. Aristotle continues,

> The reason why the movements relapse is this. The agent is itself acted upon by that on which it acts; thus that which cuts is blunted by that which is cut by it, that which heats is cooled by that which is heated by it, and in general the moving cause (except in the case of the first cause of all) does itself receive some motion in return; e.g. what pushes is itself in a way pushed again and what crushes is itself crushed again. Sometimes it is altogether more acted upon than acting, so that what is heating or cooling something else is itself cooled or heated, sometimes having produced no effect, sometimes less than it has itself received. (This question has been treated in the special discussion of action and reaction, where it is laid down in what classes of things action and reaction exist.) Now that which is acted on escapes and is not mastered, either through deficiency of power in the concocting and moving agent or because what should be concocted and formed into distinct parts is too cold and in too great quantity. Thus the moving agent, mastering it in one part but not in another, makes the embryo in formation to be multiform . . . (πολύμορφον).[40]

Aristotle's explanation for why the relapses occur explains how the semen can be affected by the menses, how that which is actualising can be affected by that which it actualises, how that which acts is affected by that which it acts upon especially in the case where what is acted upon shares the potentiality of that which acts upon it. In these cases, as in the menses' relation to the semen, what is acted upon is already on the way towards being like that which acts upon it. That which is acted upon escapes the work of what acts upon it when what acts upon it is not sufficiently capable of acting, or when what is acted upon is too cold or too much. When this happens, and it can happen in

---

[39] Witt, 'Form, Reproduction, and Inherited Characteristics', pp. 56–7.
[40] *GA* 768b15–29.

one way and not in another, the embryo that results can be actualised in some ways and not in other ways, and in this sense, is 'multiform'. Connell argues that the actual movements of the father are the current potential movements, while the potential movements of the mother are the future potential movements to show how the well-concocted mixture of male and female produces the actuality of the current potential, but it could fall short of being well-concocted and then the future potentiality would come to be.[41]

Against the view that because the movements are individual, their effects are not formal, if form is individual and formed from a material similar to the menses which itself shares a degree of what made form in semen form, then the individuality does not compromise the formal aspect of those movements where the female contribution can approach the influence of form and the male contribution can be overcome by that power in the female.[42] If form is species form, then form is unable to explain the individual differences a child inherits from the father. If form is individual form and the source of inherited traits, then it would come only from the father and so not explain differences from the mother unless all the differences are included potentially in the father. If inherited differences are not formal but material, and the male contributes nothing material, then the inherited differences cannot come from the father. In the process of generating the offspring, what is formed is a being who either can or cannot bring the source of life into another. That process occurs depending on how well the semen can concoct the menses. If it concocts it to a certain degree, it produces life, but it has to concoct it to another degree for that life to be capable of producing life in another. This process of concoction would seem to be one process, not two separate ones, where first life is formed and then the capacity to form life in another is formed. If sex differentiation is part of the first process, the process of forming resemblances would seem to be so as well, since Aristotle describes the process of resemblances as first being male or female and then resembling the parent who is male or female. The explanation that divides between the work of form and the work of the individuals based on a two-step process repeats the division between transcendent form and individually existing natural thing that the Neoplatonists develop out of Plato. Aristotle's accounts of both sexual differentiation and inherited resemblances point to how form is affecting and being affected by material.

---

[41] Connell, *Aristotle on Female Animals*, pp. 314–15.
[42] Connell argues that the movements can be neither formal nor material since they come from both parents in ibid. p. 321.

Contradiction and Contrariety

The arguments for the Möbius strip nature of the relationship between form and matter in generation for Aristotle show the inadequacy of both the one-sex and two-sex models for thinking about the relationship of form to matter, and by extension of male to female, in Aristotle. The one-sex model makes the one the lack of the other, and therefore the opposite: the one is form and the other is privation of form. The two-sex model makes the one wholly other than the other. While on the second model, the other is not defined as not the first, it being wholly other makes each seem as if it is what it is by being wholly other than the other. On the first model, the female is the inverted male, the lack of male capacity. On the second, where form and matter are perhaps an easier way to think about this relationship, each seem to have their own capacities, but form has a positive capacity and material only the capacity to be affected by form. On both accounts, form and matter are opposites, form is definition and material is what exists to be defined, a separate capacity, but one that still follows from lacking form. This argument has shown that material has its own character and that natural form functions through material.

This section considers the numerous passages in which Aristotle describes the relationship of the male and female as contraries or opposites to question the extent to which this relation is one of opposition between a positive pole and a negative one, where only one capacity would seem to truly be relevant. I argue that the relationship of matter as opposed to form but not as privation of form shows how Aristotle in struggling to explain this relationship describes it in ways more akin to the Möbius strip than either the relation of contraries or contradictories. It is not one of contradictories because the relation is not one of an excluded middle, but it is also not simply opposed but irreducibly different. The material side exists as less dependent on form as that which in a sense underlies and remains and the form is more dependent on matter to be what it is.

Aristotle relies on contraries to explain how substantial generation is possible – the changes come from the poles of the contrary. The contraries are the x and not-x that cannot be characterised of the same thing in the same way that Aristotle describes in *Metaphysics* IV.4 in the context of the law of non-contradiction. If what comes to be comes from its contrary, then the contrary pole somehow must include the not in what it is. To avoid this problem, Aristotle, in a move not unlike Plato's view that contraries can endure in a thing by participating in contrary Forms, posits some existing substratum that takes on the contraries. For Aristotle, the contraries cannot be in the substratum in the same way at the same time. By contrast to Plato, Aristotle argues that the form that is the contrary (its contrary is the privation of that form) is the essence of what emerges as the natural substance. Because form wholly characterises substance, that which comes to be cannot be both the form and not the form, except in the sense of potentially and actually. The essence characterises the natural substance and yet at the same time needs something other than itself or its privation in order for the generation to be possible. This other thing needs to be capable of being transformed by the form to become natural substance while still having its own capacities. If it does not have its own capacities then it cannot be other than the privation of form, so then form would just come from contraries, and then the contrary would include its contrary and so nothing could be differentiated from anything else.

Material, Aristotle tells us, is a third principle to the two principles of the form and privation and as such is neither the form nor its privation, but something otherwise. In *Physics* I.7, Aristotle explains generation as something that happens between contraries, 'since it is impossible for the contraries to be acted on by each other. But this difficulty also is solved by the fact that what underlies is different from the contraries for it is itself not a contrary (ἐναντίον).'[43] In *Physics* I.9, Aristotle clarifies how the material as what underlies differs from privation: 'Now we distinguish matter and privation, and hold that one of these, namely the matter, accidentally is not, while the privation in its own nature is not; and that the matter is nearly, in a sense is, substance, while the privation in no sense is.'[44] When Aristotle says that material is accidentally what is not, the point is not that material has no character of its own, but that the proximate material in generation is of some other character than it will be once generation has occurred. The material is not by definition not-what-comes-to-be, while the privation of form is. Aristotle refines his definition of matter in this same chapter: 'For my definition of matter is just this – the primary substratum of each thing, from which it comes to be, and which persists in the result, not accidentally.'[45] If material is the primary substratum as substance is the substratum in *Metaphysics* VII.3, then it is substratum as its own separable 'this' or τόδε τι. Material is never *per se* not, as the privation is. Material from which natural generation occurs, whether that is viewed as the seed or the menses, is already proximate material, worked up for the sake of

---

[43] *Phys.* 190b32–4.
[44] *Phys.* 192a4–6. Gill argues that as the substratum material has an identity of its own and it is the subject of generation (not merely of destruction) with reference to *Physics* I.7 in *Aristotle on Substance: The Paradox of Unity*.
[45] *Phys.* 192a31–3.

being acted upon by sperm. This material is transformed through this process, but is capable of being so transformed because of the elemental capacities that are the bedrock of matter. This bedrock capacity is generic material that carries all the power of the elements, which is why form continues to work on it because the generic material continues to move toward its own ends. Like substance, the elements can be on their own terms. But material does not remain as substratum in substance, which becomes its own substratum as a separable this. If material remained as a substratum that was separable and not transformed in generation, then substance would be multiple and not unified, and therefore, not a separable this. Specifically, in generation, the material contribution has its own capacities other than form without being not-being itself.

In light of this distinction between form as contraries and material as something other than those contraries, it is surprising to see Aristotle describe male and female as contraries in *Generation of Animals*. At *Generation of Animals* I.18, Aristotle maintains that 'all the products of semen come into being from contraries (ἐξ ἐναντίων), since coming into being from contraries is also a natural process, for some animals do so, i.e. from male and female'.[46] Here male and female are contrary principles that come together to form the offspring. In generation, the male and female contributions are opposed around the capacity to bring life. But the relationship is not one in which the one is destroyed by the presence of the other, nor where the two are mixed to form a third, except in a sense. Three books later, in *Generation of Animals* IV.1, Aristotle writes that the two sexes literally are opposed – τοῦτον ἀντίκειται – in their ability to 'reduce the residual secretion to a pure form'. Not only are the sexes opposed, but the force of one contribution or its failure can turn the offspring into a being capable of departing a similar contribution or falling short of that capacity:

> [. . . W]e must understand besides this that, if it is true that when a thing perishes it becomes the opposite of what it was, it is necessary also that what is not under the sway of that which made it must change into its opposite. After these premises it will perhaps be now clearer for what reason one embryo becomes female and another male. For when the first principle does not bear sway and cannot concoct the nourishment through lack of heat nor bring it into its proper form, but is defeated in this respect, then must the material change into its opposite (τοὐναντίον). Now the female is opposite (ἐναντίον) to the male, and that in so far as the one is female and the other male.[47]

[46] *GA* 724b7–10.
[47] *GA* 766a13–22.

What is not under the sway of that which made it changes into the opposite of that which made it. The material that is worked on turns into the opposite of what the proper form attempts to bring about. It is after this passage that Aristotle notes that certain parts are principles of the whole body so that when they change the whole body does,[48] which leads Aristotle to pose a number of conditions, which if met, would explain when and how the heart and blood are formed:

> ... [T]he male is a principle and a cause, and the male is such in virtue of a certain capacity and the female is such in virtue of an incapacity, and [...] the definition of the capacity and of the incapacity is ability or inability to concoct the nourishment in its ultimate stage [...] and [...] the cause of this capacity is in the first principle and in the part which contains the principle of natural heat ...[49]

If we accept these conditionals, the male and female are contrary principles or opposites of a kind because of a presence or absence of a capacity to concoct, a capacity that comes about through having sufficient appropriate heat to achieve the right level of concoction. An opposition is set up between the male and the female in reproduction, an opposition of contraries that is based on a capacity or incapacity, which is to say on a form and privation. But the male and female animals are not a form and a lack. For one, the male and female contributions do not develop in opposition but from the same process, both having this natural heat within themselves, both concocting nutrients to further serve their living. The contributions that serve reproduction are not opposites, but different capacities in a process of generation where the female contribution's failure to achieve the same degree of heat in a preliminary process of concoction makes it 'as it were' deformed. It is deformed in the sense of achieving the capacity of bringing life into the offspring, but for the sake of its own purposes, perfectly capable of achieving them, and sometimes even of achieving the capacity designated to the male that of producing another like it, though, as with the male, requiring the contribution of the male in order to do so. The female is the privation of this particular capacity, but having its own capacity, able to become the living organism when male form acts on female matter.[50]

---

[48] GA 766a23–9.
[49] GA 766a30–3, 35.
[50] James Bogen argues that Aristotle's use of the term 'complete privation' at *Metaphysics* 1055a33ff. suggests that there can be partial privation in 'Change and Contrariety in Aristotle', pp. 13–14. See also Gill, who argues from *Physics* I.7 and *Generation of Animals* I.18 that material as the ὑποκείμενον in generation has its own proper identity and it remains at work when formed by the form and seed in *Aristotle on Substance*, pp. 106–7.

Generation brings together these two capacities while sexual differentiation is the process of distinguishing the capacity of the embryo to bring forth another, but it still does not establish the female as by definition privative. In sexual differentiation, one contribution can fail to assert its capacity even though its capacity would seem by definition to be the capacity to assert this capacity. The semen, capable of bringing life into another by heating it to the point where it not only lives but is capable of making another live, can fail to have sufficient movement or because of too much material can fail to properly concoct it all so that the embryo does not have the capacity to produce semen and therefore this same capacity. The opposition between the male and female principles in this process moves along a path from the work of one to the other. This movement can be either because of what semen fails to do or how menses overpowers it.

The shift from male to female, a shift that establishes the contrary between the formal and material contributions, occurs between opposites in a way that shows a difference not between form and matter but within the material body, as Aristotle describes the transition that occurs from the offspring being male to the offspring being female in *Generation of Animals* IV.3:

> Now, when anything *departs from type* (ἐξίσταται, a form of ἔκστασις), it goes not into any chance thing but into the opposite (ἀντικείμενον), and so too in generation, what isn't mastered necessarily departs from type and comes-to-be the opposite with respect to the δύναμις with respect to which the generator and mover didn't get mastery. If, then, it's *qua male*, what comes-to-be is female.[51]

Here the change is at the level of material as shown above. Within the material body there are contraries (necessary ones, which would seem to make them contradictories) – male and female – that have the capacity to form the offspring from the generative principles of form and material. These are material contraries that become formal contraries in generation. They become formal contraries through material difference, that of heat.

> Now the female is opposite to the male, and that in so far as the one is female and the other male. And since it differs in its faculty, its organ also is different, so that the embryo changes into this state. And as one part of

---

[51] GA 768a2–5. Furth translation and emphasis, *Substance, Form and Psyche*, p. 130.

first-rate importance changes, the whole system of the animal differs greatly in form along with it. This may be seen in the case of eunuchs (εὐνούχων), who, though mutilated (πηρωθέντος) in one part (μορίου) alone, depart so much from their original appearance (τῆς ἀρχᾶς μορφῆς) and approximate closely to the female form (θήλεος τὴν ἰδέαν). The reason is that some of the parts are principles (τῶν μορίων ἀρχαί), and when a principle is moved many of the parts that go along with it must change with it.[52]

In *Categories* 10, Aristotle describes two kinds of contraries: those in which one or the other contrary must belong to that of which they are contraries, as sickness or health must belong to animals' bodies and odd or even must apply to numbers and those in which neither extreme necessarily belongs to that of which they are contraries as black and white need not belong to a body and bad or good need not be predicated of men.[53] For the second kind of contrary, the one extreme can change into the other.[54] This kind Aristotle simply calls contraries. The first maintains an uncrossable distance. This kind of contrary Aristotle calls contradictories. At *Metaphysics* X, Aristotle writes that 'contradiction (ἀντίφασις) admits of no intermediate, while contraries (ἐναντίων) admit of one'.[55] Aristotle continues that while contradiction does not allow an intermediate, the change in matter is from contraries. Several chapters later, he writes, 'Since contraries (ἐναντίως) admit of an intermediate and in some cases have it, the intermediate must be composed of the contraries.'[56]

Two chapters later, Aristotle addresses gender and contrariety:

One might raise the question, why woman does not differ from man in species, female and male being contrary, and their difference being a contrariety (ἐναντώσεως); and why a female and a male animal are not different in species, though this difference belongs to animal in virtue of its own nature, and not as whiteness or blackness does; both female and male belong to it *qua* animal. This question is almost the same as the other, why one contrariety makes things different in species and another does not.[57]

---

[52] *GA* 766a21-9.
[53] *Cat.* 12a1-17.
[54] *Cat.* 13a19-21.
[55] *Meta.* 1055b1-2.
[56] *Meta.* 1057a18-19.
[57] *Meta.* 1058a29-36.

After explaining that contraries that are in formula make a difference in species, while contraries in the material do not, he concludes:

> And male and female are indeed modifications peculiar to animal, not however in virtue of its substance but in the matter, i.e. the body. This is why the same seed becomes female or male by being acted on in a certain way. We have stated, then, what it is to be other in species, and why some things differ in species and others do not.[58]

Contraries that must belong to a thing are contradictories, admitting no middle, and different in the sense of x and not x in a way that makes the contrary a formal one and therefore a species difference. Contraries in the body, contraries that are in material, seem capable of passing through one another, or of becoming the one on the way toward being the other. This distinction is what leads Marguerite Deslauriers to argue that sexual difference is non-accidental and necessarily belonging to animals, but material.[59] Since sexual difference is not part of form, it is not a species difference and so not an essential difference. While Deslauriers argues that sexual differentiation cannot be formal because it does not occur at the time of conception, a position that would seem to require a separate movement of semen and menses for sexual differentiation, even if sexual differentiation occurs at the time of conception, I maintain the process is not only determined by material but results in different material capacities for contributing material or form. The degrees of heat cause the distinction. The result is different parts that allow for different capacities of developing heat. Arguments over inherited traits reach similar impasses regarding whether the traits like sexual difference are material or formal because the contributions to drawing the distinction are themselves distinguished as formal and material. The changes that occur in sexual differentiation would seem to produce a capacity for form and a capacity for material, and yet the results and the process of changing appear very much to be material changes. If the contraries of male form and female matter are themselves contraries of form, male and female would be separate species. If the contraries of male form and female matter are contraries of matter, then the distinction between form and matter itself seems produced and rooted in material. The Möbius strip nature of Aristotle's account seems to require that the distinction is not reducible to either a formal or a material one because

---

[58] *Meta.* 1058b21–5.
[59] Deslauriers, 'Sex and Essence in Aristotle's Metaphysics and Biology'.

neither of the consequences seem tenable for Aristotle. The formal contribution cannot be the result of material difference, which itself would seem to require material to produce form; but the distinction cannot be so severe that there is no connection between them, so Aristotle explains it in material terms, that the distinction is produced by and produces material and bodily difference.

7
Craft and Other Metaphors

Craft Analogy Explains Why Generation Needs Males and Why It Happens in Females

Commentators have pointed to Aristotle's use of craft analogies to explain generation to argue that Aristotle views natural generation according to a paradigm of artifice.[1] Such a view would assign to Aristotle a notion of nature as a demiurgic force which works externally to that which is formed. This analogy contributes to the view that Aristotle has a conception of prime matter – basic stuff that can only become characterised through some external cause – and of species form, which causes individual substances at some remove. Once the case has been made against prime matter and on behalf of individual form, the question is how to understand Aristotle's regular reference to craft analogy, especially in the biological works. This chapter shows the ways that the analogy to craft is invoked to explain specific aporiai in Aristotle's account of generation rather than to assign to nature the structure of craft. Some references even point to how, following the position first laid out by Charlton, maintained by Tress and more recently pursued by Connell, the truth of the analogy to craft for Aristotle is in the ways the analogy is a disanalogy.[2] The last section

---

[1] Emanuela Bianchi argues that the craft structure is a problematic avenue toward giving a robust role to the female in Aristotle's account of nature in *The Feminine Symptom*, pp. 61, 63-4, 192-7. Supporters of prime matter in Aristotle draw heavily on the analogies to craft to show that material is without character of its own; see Friedrich Solmsen, *Aristotle's System of the Physical World*, pp. 122-3 and Solmsen, 'Nature as Craftsman in Greek Thought', pp. 489-92.

[2] William Charlton, *Physics* I and II, pp. 89-92. Sophia M. Connell, *Aristotle on Female Animals: A Study of the* Generation of Animals, pp. 123-7, 156-60. Tress argues that Aristotle's account of generation stands in contrast to modern notions of reproduction that draw on

of this chapter addresses the other images for generation Aristotle employs in *Generation of Animals* that are typically underplayed at the expense of the craft analogy.

Aristotle's first use of the craft analogy in *Generation of Animals* is in I.21, where Aristotle explains the manner in which the male contribution is in the embryo and the cause of it.

> Therefore, if we take the highest genera under which they each fall, the one being active and motive and the other passive and moved, that one thing which is produced comes from them (ἐκ τούτων) only in the sense in which (ἀλλ' ἢ οὕτως) a bed comes into being from the carpenter and the wood, or in which the ball comes into being from the wax and the form (τοῦ εἴδους). It is plain (δῆλον) then that it is not necessary that anything at all should come away from the male, and if anything does come away it does not follow (οὔτ') that this gives rise to the embryo as being in the embryo (ἐνυπάρχοντος), but only (ἀλλ') as that which imparts the motion and as the form (ὡς ἐκ κινήσαντος καὶ τοῦ εἴδους); so the medical art cures the patient.[3]

The focus of this chapter is on how the male contributes to generation and how the semen is the cause of the embryo. Aristotle makes an *a priori* argument for how generation would have to work, which he confirms in the passage that follows: 'This *a priori* argument is confirmed by the facts.'[4] Aristotle is accounting in this passage for all animal generation that occurs between the male and the female, which requires an account of insects where the female inserts something into the male. He argues that nothing necessarily would have to come away from the male, which is not to say that it does not, as he goes on to observe and explain. Aristotle is trying to understand the general structure of the work of the male that makes sense of the cases in which the male contributes motion that is not itself through some separate material, as in the case of insects, and

---

metaphors of production of artefacts in 'The Metaphysical Science of Aristotle's "Generation of Animals" and Its Feminist Critics', p. 311. Connell argues for local interpretations of the craft analogy and warns against taking it too seriously, *Aristotle on Female Animals*, pp. 159, 176n42, 212, 219. Mariska Leunissen similarly raises questions about the sense in which form is a craftsman because the craftsman sets the form from the outside, and art itself does not deliberate, while nature does adapt and work on organising itself to fulfil its end because its principle is internal to it in '"Crafting Natures": Aristotle on Animal Design'.

[3] *GA* 729b14–20.
[4] *GA* 729b21.

when the male contributes motion through some material. The umbrella structure shows that the male contributes motion and form.

Aristotle uses three metaphors – bed, wax and medicine – to explain *a priori* how the moving principle works in relationship to the passive material principle (this is the precise language of the previous several sentences).[5] The carpenter and the ball form their products without contributing any material to what is formed, that is the work of the wood and the wax, though each do the work in and through material. The carpenter works with material tools, and the movement depends on material processes, but the bed becomes bed by being organised by this movement of the carpenter. The point is that movement and not some other material makes the wood bed. The form of the wax ball is the shape rather than the shaper of the ball, but here again, the point is the form does not contribute wax but the shape of wax.

If these examples establish an efficient and formal cause as ways of thinking of how the semen causes, Aristotle concludes with another possibility for how the moving cause works, the analogy to a medical art. The medical art starts a motion of health that the patient's body takes over and actualises in order to become healthy. If the body did not already have an inclination toward health, the medical art would be imposing an external form upon it. The medical art motivates capacities toward health that are already at work in the patient's body. As the living organism strives toward health, it can be nurtured more in that direction, not in a way that is wresting it away from its natural condition, but in a way that starts it on a path toward actualising that condition. This model presents the possibility that the male contribution works on the female contribution that is already on the way to life as it has been concocted already to a certain degree, and in fact, what animates it also became capable of animating by passing through the levels of capacity that the female contribution has and going further.

The craft analogy returns in the following chapter to address the reason generation must occur in the female because she is the source of material. While the previous use of the analogy explains that the male does not contribute material but movement in generation, this use of the analogy establishes the need for the male and female to be in proximity in the process. If the male did not have to contribute anything, as the previous passage suggests, the proximity would be unnecessary. As Aristotle explains, the embryo grows in the female because just as the carpenter and potter must remain close to the material being formed, the forming movement must remain connected to the

---

[5] GA 729b5–14.

material. Aristotle then turns to the final example of architecture to affirm that architecture – the work of building – is in the edifices built.

> [... B]ut the female receives within herself the share contributed by both, because in the female is the material from which (ἐξ ἧς) is made the resulting product (δημιουργούμενον). Not only must the mass of material exist there from which the embryo is formed in the first instance, but further material must constantly be added that it may increase in size. Therefore (ὥστ') the birth must (ἀνάγκη) take place in the female. For the carpenter must keep in close connexion with (πρὸς) his timber and the potter with (πρὸς) his clay, and generally all workmanship and the ultimate movement (ἡ κίνησις ἡ ἐσχάτη) ~~imparted to matter~~ must be connected (πρὸς) with the material concerned, as, for instance, architecture is *in* (ἐν) the buildings it makes.
>
> From these considerations, we may also gather how it is that the male (ὁ ἄρρεν) contributes (συμβάλλεται) to generation (πρὸς τὴν γένεσιν). The male does not emit (προΐεται) semen (σπέρμα) at all in some animals, and where he does (προΐεται) this is no part of the resulting embryo; just so no material part comes from the carpenter to the material, i.e. the wood in which he works (οὐδ' ἀπὸ τοῦ τέκτονος πρὸς τὴν τῶν ξύλων ὕλην), nor does any part (οὔτε μόριον) of the carpenter's art exist within what he makes, but the shape (ἡ μορφὴ) and the form (τὸ εἶδος) are imparted from him to the material (ἐν τῇ ὕλῃ) by means of the motion he sets up (διὰ τῆς κινήσεως). It is his hands that move his tools, his tools that move the material; it is his knowledge of his art, and his soul, in which is the form, that move his hands or any other part of him with a motion of some definite kind, a motion varying with the varying nature of the object made (καὶ ἡ μὲν ψυχὴ ἐν ᾗ τὸ εἶδος καὶ ἐπιστήμη κινοῦσι τὰς χεῖρας ἤ τι μόριον ἕτερον ποιὰν τινα κίνησιν, ἑτέραν μὲν ἀφ' ὧν τὸ γιγνόμενον ἕτερον, τὴν αὐτὴν δὲ ἀφ' ὧν τὸ αὐτο, αἱ δὲ χεῖρες τὰ ὄργανα τὰ δ' ὄργανα τὴν ὕλην). In like manner, in the male of those animals which emit semen, nature uses the semen as a tool (καὶ ἡ φύσις ἐν τῷ ἄρρενι τῶν σπέρμα προϊεμένων κρῆται τῷ σπέρματι ὡς ὀργάνῳ) and as possessing motion in actuality, just as tools are used in the products of any art, for in them lies in a certain sense the motion of the art (ἡ κίνησις τῆς τέχνης). Such, then, is the way in which these males contribute to generation. But when the male does not emit semen (προΐεται), but the female inserts (ἐναφίησι) some part of herself into the male, this is parallel (ὅμοιον) to a case in which a man should carry the material (ὕλην κομίσειέ) to the workman (πρὸς τὸν δημιουργόν). For by reason of weakness in such males nature

is not able to do anything by any secondary means, but the movements imparted to the material are scarcely strong enough when nature itself watches over them. Thus here nature resembles a modeler in clay rather than a carpenter, for she does not touch the work she is forming by means of tools, but with her own hands.[6]

The first paragraph of this passage focuses on why generation takes place in the female. The second focuses on how, in light of generation taking place in the female, the male contributes to generation. In the first paragraph, Aristotle explains that birth takes place in the female because the movement must be close to the material. In this passage, Aristotle explains that the craftsperson must work in what is being built, but he does not give an indication that the employment of the analogy is supposed to be carried further than this point in this passage. The reference to architecture at the end of this paragraph points to how the analogy is both an analogy and disanalogy. Architecture both is and is not in the building it makes. Architecture is in the building because the work of making the building must be in the building, which is consistent with Aristotle's account in *Physics* III that motion is in the movable, as the work of teaching and learning coincides in the student, and in *De Anima* II.11 that the actualising occurs in what is potential.[7] In natural things, the natural source that does the work in what is moved, remains in and becomes internal to what is formed. The natural form stands in contrast to artefacts in this sense because architecture, insofar as it is in the mind of the architect, remains external to the building. In contrast to teaching, Aristotle's example in *Physics* III, the building does not take on the know-how of the builder when it is actualised, it takes on the form that the builder has of how the building should be. The teacher actualises the learner in a way that makes the learner like the teacher, but the building does not become like the builder, carrying the same capacities, but like the form that the builder imposes. When Aristotle says that architecture is in the buildings, he points to the part of the analogy that emphasises that the work of form is in what is formed, but not the whole of the analogy. More evidence that Aristotle is not being precise about the analogy is that he says that the architecture is in the building to explain how it is in the mother, and yet that which is being formed is more the embryo than the mother. Architecture as an art occurs in the building, but the work that comprises building remains outside. Aristotle confirms this point in the next

---

[6] *GA* 730a35–730b31.
[7] *Phys.* 202b3–8, 17–21; see also *Meta.* 1046a19–28. *De An.* 426a6–11.

paragraph when he argues that no part of the carpenter's art exists in what the carpenter makes. By contrast in natural generation, the form that comes from outside is internalised in the embryo so that the organising work is now what it is to be the embryo. The craft analogy does not quite give Aristotle the language and concepts that he needs, but it does explain that development happens in the material that is formed, and this placement is why it occurs in the mother.

Aristotle explains the work of semen in comparison with the carpenter, who brings neither material nor a part of the art, but the shape or form, into what is made. The analogy allows Aristotle to show that there is a principle of generation that is not reducible to material. But the analogy falls short because it cannot explain how the external ordering movement becomes internal movement. Against predecessors who see generation as a mixing of a male contribution and a female contribution which equally contribute material, Aristotle offers the craft analogy to explain how the form works on the material without contributing material, continuing his effort to offer a generic account of animal generation of male and female. On the structure of craft, the carpenter acts through material in the form of tool without imparting material. In the situation of the modeller of clay, the modeller does not leave some material behind, but neither does the modeller work with tools. In this case, nature 'does not touch the work she is forming by means of tools, but with her own hands'. In this case, Aristotle says that when there is nothing imparted from the male, but the female 'inserts some part of herself into the male', nature is working directly unmediated by tools. Aristotle's effort to present a consistent model ends up making the apparently passive form of the male work directly on the female contribution, though it is the female that is working upon the male in this scenario.

Aristotle needs to explain how the motions of the semen can be transformative so that they can bring forth life in the material contributed by the mother. The craft analogy explains generation in terms of spatial motion, but spatial motion fails to capture the transformative work of form through semen.[8] Aristotle refers to the way that the material capacities of semen cause motion in such a way that transforms what it moves rather that merely moves it across space. What is remarkable about this transformation is that it occurs through the material capacities of the elemental forces that constitute semen. On this basis, the craft analogy can be taken to explain the particulars of this situation – how the male can act without having something left behind, much

---

[8] See Connell, *Aristotle on Female Animals*, p. 168.

as the tool of the craftsperson does – but it cannot do the work of explaining how semen is a transformative principle of motion in generation.

Connell concludes from the fact that some males contribute to generation without emitting something that nature does not need a mediating tool: 'It, therefore, does not require a chain of physical reactions to bring about its goals.'[9] Nature works directly when nothing is emitted, but when semen is emitted, nature works through some specific material whereby it does its work. This explanation makes nature work more directly and unmediated in lower animals, which follows the Neoplatonist view of how form works. Connell argues from the example of animals where the male does not emit semen that Aristotle does not offer material explanations for the source of the nutritive soul, rather nature is itself the external principle to the process of generation. For her, the analogy to the carpenter in this passage disallows the material aspect of semen from explaining soul.[10] Yet another way of reading the analogy to the carpenter is that semen is a tool that needs specific material characteristics, just as the saw and the carpenter's hands do, in order to effectuate the final product. The carpenter's hands move by means of the carpenter's knowledge of the form. The question is what corresponds in nature to the knowledge in the mind of the carpenter.

Unlike the saw and the carpenter's hands, semen carries its capacities for motion within it – in the work of heat – to transform that which it works upon in very material ways that also result immanently in a change of form. The natural form transforms and semen would seem to be the vehicle through which that transformation occurs. If semen is strictly a tool in the sense that the saw or even the hand is a tool, then it would act to change the shape and place of the material, but not to transform it. If the analogy to art is to stand, form must have some way of existing externally to its tool. But there are important senses in which this externality of form does not seem possible for Aristotle.

At *GA* I.23, Aristotle describes nature as an intelligent workman (εὐλόγος ἡ φύσις δημιουργεῖ).[11] He proceeds to explain how the division of sexes into two in many animals contributes to the fulfilment of their end in participating in a kind of knowledge. Aristotle does not argue that nature is a workman imposing form on material, but rather argues that nature is an organising principle. This principle need not be considered external to what is formed.

---

[9] Ibid. p. 212.
[10] Ibid. p. 183.
[11] *GA* 731a25.

Craft Analogy Explains How Semen Actualises Form

Aristotle moves toward the specifics of how semen works in *Generation of Animals* II.1, where he likens semen to automatic puppets:

> As, then, in these automatic puppets the external force moves the parts in a certain sense (not by touching any part at the moment, but by having touched one previously), in like manner also that from which the semen comes, or in other words that which made the semen, sets up the movement in the embryo and makes the parts of it by having first touched something though not continuing to touch it. In a way it is the innate motion that does this, as the act of building builds the house. Plainly, then, while there is something which makes the parts, this does not exist as a definite object, nor does it exist in the semen as a complete part.[12]

Aristotle is trying to explain how the movement can come from the father and yet be otherwise than the father.[13] The automatic puppet explains how the movement can belong to what moves and still be caused by something else. Like the automatic puppet's movements, the movement in the semen comes from the parent who moves not by presently moving it, as a carpenter might move a tool, but by having previously been in contact with it and setting it in motion. The material capacities of the semen enable it to continue to have the motion imparted from the father at work within it in such a way that it can also set up this movement in the embryo. In this sense, through the semen, the father remotely moves the maternal material to become the embryo. As the automatic puppet, the movement of the semen is external to it and to the embryo.

Invoking a different analogy, Aristotle says it is the innate motion in the semen that moves the embryo, 'as the act of building builds the house'. The building occurs in the activity. Above we saw that architecture occurs in the house and now we see that the activity of building is what builds the house. The point is not that the activity comes from outside but that the activity is what actualises that in which the activity occurs. The activity of the semen forms the embryo. The male parent moves transitively through the semen to animate the menses through its motion. Aristotle writes, 'Now the semen is of such a nature, and has in it such a principle of motion, that when the motion is ceasing each of the parts comes into being, and that as a

---

[12] *GA* 734b12–18.
[13] Following Connell, *Aristotle on Female Animals*, p. 159.

part having life or soul.'[14] The semen's work is this motion, which is why it ceases to be once the part having life or soul (the heart) is formed. The semen both has and sets up the 'principle of movement' that defines nature, as in *Metaphysics* IX.7, where Aristotle writes, 'For nature also is in the same genus as potentiality; for it is a principle of movement – not, however, in something else but in the thing itself qua itself. To all such potentiality, then, actuality is prior both in formula and in substance.'[15]

Aristotle proceeds by recalling the specifics for how semen transforms the material from what he established in *Meteorology*. There Aristotle shows how heat and cold work on dry and moist to form homogeneous parts, but to form nonhomogeneous parts and ultimately the organism, more than hot or cold is needed. As Aristotle writes, once more invoking a metaphor of craft: 'What makes them a sword is the movement of the tools employed, this movement containing the principle of this art.'[16] Heat is inadequate to form the sword, not because heat cannot do the work, but because heat needs the principles contained in the movement, principles for organising and forming in order that the thing might be formed to fulfil its function. The movement makes the sword because the movement implies a purpose, making something sharp for fighting in war. The analogy need not require that the principles are external, but that ordering principles accompany the heat. The heat of the semen has a capacity as natural heat and πνεῦμα to move toward life. The semen expresses its power as form by moving the mother's material in such a way that shapes it into animal life. The directedness of heat toward life is the power of form. That which it acts upon has also become what it is through the directedness of heat that achieves its end by becoming that which a little bit more heat can act upon to animate. Aristotle insists that the heat must be directed through this analogy to the heat that makes the sword. The directed heat is what forms the sword. For Aristotle, vital heat is this directed heat, the heat whose power is life.

In *Generation of Animals* II.4, Aristotle again explains as he did in I.21 and I.22 that not all males have a generative residue while all females do because the female provides the material and the male fashions it. As Aristotle writes, 'Thus while it is necessary for the female to provide a body and a material mass, it is not necessary for the male, because it is not within what is produced that the tools or the maker must exist.'[17] The claim is that the male does not need to offer material, while the female must. If the male offered material it

---

[14] GA 734b22–4.
[15] Meta. 1049b7–11.
[16] GA 735a1–2.
[17] GA 738b22–5.

would be as tools, but that does not happen because the tools do not need to exist in what is produced. The semen does not continue to exist as the material in what is produced because its work is to act to bring forth form.

Aristotle proceeds to discuss the technical workings of conception. The female discharge is outside the uterus, where due to its heat, it attracts the semen. Then the female material is fixed by the male as rennet acts on milk, a metaphor considered in further detail below. The rest of the chapter is about the growth that occurs after the embryo is initially formed. Aristotle compares the way the embryo acts as it grows to seeds in plants which 'contain the first principle of growth in themselves'.[18] Aristotle will say later in this chapter that the female contribution contains all the parts potentially; in this passage, he makes the same claim about the embryo. The heart is the first part that is actualised and the heart becomes the first principle of growth in the embryo.

In the context of this treatment of growth, Aristotle addresses the question of the extent to which the embryo grows from what is external to it and the extent it grows from what is internal to it. Since it is 'already potentially an animal but an imperfect one' it cannot be self-nourishing, but draws nutriment from the mother as the plant does from the earth.[19] The heart forms a vessel toward the mother and the umbilical cord is formed.[20] Asking how the heart can contain blood through which it can form the vessel to attract the nourishment of blood, Aristotle writes, 'Perhaps it is not true that all of it comes from outside.'[21] When it is formed, the embryo already has some way of nourishing itself. The embryo grows from drawing nutriment through the umbilical cord, the plant through its root and the animal from the nutriment it draws into itself. In each case, something is drawn in from outside. But the living thing orders it toward its end of living. The nutritive soul grows through this nourishment, but the growth occurs because the potential is already in the female contribution, which is akin to the nutriment. As Aristotle writes, 'The real cause of why each of them comes into being is that the residue of the female is potentially such as the animal is naturally, and all the parts are potentially present in it, but not actually.'[22] The residue of the female is potentially animal, which explains how it is both potentially alive and the source of continued nutrition, a capacity the embryo takes over from the female contribution, which is concocted blood, which develops from

---

[18] *GA* 739b35.
[19] *GA* 740a24–6.
[20] *GA* 740a29–31.
[21] *GA* 740b4–5.
[22] *GA* 740b19–20.

nutriment, and is potentially alive. It is at this point that Aristotle employs the craft analogy:

> The female, then, provides matter, the male the principle of motion. And as the products of art (τὰ ὑπὼ τῆς τέχνης) are made by means of the tools of the artist (διὰ τῶν ὀργάνων), or to put it more truly by means of their movement (διὰ τῆς κινήσεως αὐτῶν), and this is the activity of the art (ἡ ἐνεργεια τῆς τέχνης), and the art is the form (ἡ δὲ τέχνη μορφὴ) of what is made (τῶν γινομένων) in something else, so is it with the power of the nutritive soul. As later on in the case of mature animals and plants this soul causes (ποιεῖ) growth from the nutriment, using heat and cold as its tools (for in these is the movement of the soul and each comes into being in accordance with a certain formula), so also from the beginning does it form (συνίστησι) the product of nature (τὸ φύσει γιγνόμενον). For the material by which this latter (ἡ αὐτή) grows is the same as that from which it is constituted (συνίσταται) at first; consequently also the power which acts upon it is identical with that at the beginning (but greater than it (αὕτη)); thus if it (αὕτη) is the nutritive soul, it (αὕτη) is also the generative soul, and this is the nature of every organism, existing in all animals and plants.[23]

Aristotle analogises between the ways products of art are made by the movement of the tools of the artist and the way nutritive soul is formed: as the form of what is made in something else. Aristotle explains with reference to the work of soul in mature animals and plants. As in these mature organisms, the embryo uses movement from its tools – heat and cold – because the movement of the soul is in the heat and the cold. Aristotle equates the material that originally forms the embryo to the material by which it grows, which would be the female contribution, which as we have seen is in one sense internal to the embryo and in another sense external. On the basis of establishing the equivalence of the originating material to the growing material, Aristotle is able to equate the originating form to the continuing form. The originating form was in semen, but the continuing form is in the embryo as the nutritive soul. The originating form in the semen must be the cause of this substance. It cannot also be the cause of the substance from which it comes, since substance is the essence and the essence is causal. The work that differentiates the form of the father from the form of the offspring must occur in the semen, which, through its organised movements enabled by the material that constitutes it,

---

[23] GA 740b24–741a2.

becomes the source of life for the new offspring. The material that forms the embryo and the material that continues the embryo is *the same material*, concocted material nutriment, while the form that originates coming from the semen of the father is the same in likeness and yet an independent nutritive soul that actualises the embryo.

The nutritive soul works as art does through its tools – for the nutritive soul, those tools are heat and cold. Heat and cold transform nutriment into what comes to be in something else (Platt translates τῶν γινομένων as 'what is made', but the Greek is the participle of the word for becoming, so a more precise translation would be 'what comes to be' or 'what is born'). The power of the nutritive soul is the most basic power of what comes to be in itself. Aristotle draws a parallel between how the nutritive soul works in the initial stage of generation when it first forms the natural being to how the nutritive soul directs growth. Similarly, the power that acts to form is akin to the power that acts once formed, the internalised soul. Both were acting through a tool and are in some sense equal to the work of the tool.

Heat and cold are described as tools, τῶν ὀργάνων, which implies a power that needs something other than the power to put it to work. But on Aristotle's account of how semen works to actualise soul, the tool is the work of that power. Art needs some tool to organise productive forces like heat to make what comes into being. Nature works like this, in a sense, and yet the tool that uses heat, the soul, is in the heat being of a certain degree in the form of semen. Individual form works through the materiality of elemental forces on material that has become appropriate for form through the work of elemental forces that give it the specific capacities that characterise it. On this account, the tools of form are material as the tools of soul are the body. Not only are they material, but unlike the tool of the carpenter, they are the powers that characterise that which they work upon already. Both the semen and the menses have the same origin in the nourishment, which is then differentiated from one another on the basis of degrees of heat. Semen brings soul to menses by bringing the greater heat that the menses could not work up on its own so that it can then be actualised as the living embryo. The passage explains how the nutritive soul is caused through a movement just as the artefact is formed through a movement. The reference to the tool affirms the directedness of the movement, but that directedness in natural things is internal to the tool, so that material is doing the work. This movement allows for a continuity to be maintained between the generative material and form, which come from the mother and father but are not strictly speaking the form or material of the parents, but potentially the form and material of the new offspring, and the generated form and material, which are actually of the new organism.

Aristotle uses the reference to craft in *Generation of Animals* II.6 to speak to the way that the material that forms natural things has its own character, '[T]he carpenter would not make a box except out of wood, nor will a box be made out of wood without the carpenter.'[24] This use of the craft metaphor points to the codependence of form and material, where the carpenter needs material of a certain character – even the character that makes it elemental material at its most basic level – and the material depends on a certain form (either in the mind of the carpenter or in the movement of the semen) to achieve a certain form.

The relationship between the soul or form and the specific parts that do the work of forming is raised again in *Generation of Animals* III.11 in the context of the generation of testacea. Testacea are a difficult case because some of them form spontaneously and some are formed by emitting a substance for generation. Aristotle explains this diversity with reference to the diversity of generative possibilities in plants. In each case offspring are formed from some superfluous residue through a process of concoction of a mixture with rain-water that undergoes a certain degree of putrefaction.[25] Explaining this process, Aristotle writes:

> For nothing comes into being out of the whole of anything (ἐκ παντός), any more than in the products of art (ὑπο τῆς τέχνης δημιουργουμένοις); if it did art would have nothing to do, but as it is in the one case art removes the useless material (ἀχρήστων), in the other nature does so. Animals and plants come into being in earth and in liquid because there is water in earth, and air in water, and in all air is vital heat (θερμότητα ψυχικήν), so that in a sense all things are full of soul (πάντα ψυχῆς εἶναι πλήρη). Therefore living things form (συνίσταται) quickly whenever this air and vital heat are enclosed in anything. When they are so enclosed, the corporeal (σωματικῶν) liquids being heated, there arises as it were a frothy bubble. Whether what is forming is to be more or less honourable in kind depends on the embracing of the vital principle (τῆς ἀρχῆς τῆς ψυχικῆς); this again depends on the medium in which the generation takes place (οἱ τόποι αἴτιοι) and the material which is included (τὸ σῶμα τὸ περιλαμβανόμενον). Now in the sea the earthy matter (τὸ γεῶδες) is present in large quantities, and consequently the testaceous animals are formed (συστάσεως) from a concretion of this kind, the earthy matter (τοῦ γεώδους) hardening round them and solidifying in the same manner as bones and horns (for these cannot be melted by fire), and the body which contains the life (τοῦ τὴν ζωὴν ἔχοντος σώματος) being included within it.[26]

[24] GA 743a25-6.
[25] GA 761b24-762a8.
[26] GA 762a16-31.

This passage hearkens back to earlier discussions of whether the semen is drawn from the whole body (the theory of pansomatism). Aristotle furthers his critique of pansomatism in this passage by saying that nothing generates out of the whole of anything in nature even as in art. Both nature and art work by removing what is useless from what is useful. Elsewhere Aristotle describes the residue from both male and female that contributes to generation as useful.[27] Natural generation, like artificial generation, puts to work what is useful and sloughs off what is not. The analogy to craft in this passage makes of Aristotle's account of natural generation a strikingly material account. Animals and plants, Aristotle explains, come to be in the elements, which themselves are the source of life. Animals come to be in air and in earth – all air contains vital heat, which is the way that all things are full of soul. And earth contains water which contains air, which means vital heat and thus, soul. The enclosure of liquid produces a heat, which produces a foaminess, which leads to life. In testacea, Aristotle explains, nature works as art does by distinguishing from the material that vital heat will work on and the material that it will not work on by hardening and softening to achieve a certain degree of hardness. Vital heat or the vital principle arises from these liquids when conditions warrant it to enable the distinguishing between the earth that becomes the mussel, for example, and the earth that remains the floor of the sea. Here again, Aristotle establishes a principle that does the ordering and actualising work that is still in some way related to the material potentiality of what is in play.

In *Generation of Animals* V, Aristotle has been considering how differences in material result in different outcomes. In *Generation of Animals* V.8, the very last chapter of the book, Aristotle returns to the question of how nature is like art, considering again how nature uses instruments as art does:

So it is reasonable (εἰκὸς) that nature should perform most of her operations (ἐργάζεσθαι τὰ πολλὰ) using breath (τῷ πνεύματι) as an instrument (ὀργάνῳ), for as some instruments serve many uses in the arts, e.g. the hammer and anvil in the smith's art, so does breath (οὕτως καὶ τὸ πνεῦμα) in things formed by nature (ἐν τοῖς φύσει συνεστῶσιν). But to say that necessity is the cause is much as if we should think that the water has been drawn of from a dropsical patient on account of the lancet alone, not on account of health, for the sake of which (οὗ ἕνεκα) the lancet made the incision.[28]

Aristotle explains the work of material such as breath with reference to tools – the hammer and the anvil – to argue that the material does work for

[27] GA 724b22–726a6.
[28] GA 789b8–15.

the sake of actualising the form. In this passage once again the reference to the medical art and health further explains and complicates the craft analogy. Breath is a tool as the hammer and anvil are tools. The father is the efficient cause of the breath and air in the semen in a sense, being the source of heat that works them to the point where they can do their work. But the father is not a proximate source of heat, remaining there heating in order for breath and air to work in semen. The potential of soul has already become the power of the semen. Air and breath do not have efficient causes external to that in which they work in the same way a hammer and anvil do. Like the hammer and anvil, breath does this work in order to achieve life in the living organism, to actualise the capacity of the form.

The relation between the medical art and health explains the way the tool works on behalf of form for the sake of the function the form accomplishes. Aristotle explains that the lancet does the work in bringing the dropsical patient to health. Dropsy or edema is a condition in which fluid collects in cavities or tissues causing swelling. The lancet draws off the excess fluid to reduce the swelling, because it is a direct cause of the swelling. When the water is drawn out of the body, the body can return to health. The lancet is the cause of the water being drawn off, but health is the reason or purpose for the sake of which the doctor uses the lancet. The purpose of the lancet is to draw off water, just as the purpose of breath is to make the material hot and foamy. The drawing off of water occurs for the sake of health, just as the heating occurs to animate the material.

The soul itself can be described as the tool of tools, much like the hand. The soul is a tool for life, the tool that actualises all that is required for the body to be a living body. Aristotle also describes the body as the tool of the soul. As the tool of the soul, it is the organ through which soul is manifested as soul. The image of the tool can be a way to think of how a thing is manifested in the world, both on the side of what we think of as material (body) and as form (soul). Soul animates, body manifests. On these terms, semen as a tool is that which does the work of soul. Semen has a kind of independence from the father: the motion that characterises it is started by the father but also due to the material composition of the semen so that it continues to move in a way that becomes independent from the motion begun by the father. This independence allows semen to be the potential for a new individual existing thing. Semen also has a kind of dependence on material, both on the material that allows it to be potentially alive and on the material of the female that it activates, material that is very much like it except not as worked up by vital heat. The craft analogy and the use of the image of tool do not require that form be understood as species form and semen itself as the initial working of individual form be its tool. Aristotle reaches for analogies such as tool to explain how semen can become independent from the father and still do the work begun in the form of the father. He similarly uses this analogy to explain the way that soul and body are interdependent on one another, without either being external to the other and working through it as some accounts of the tool image suggest. Aristotle's use of the craft analogies does not point to some more profound reduction of natural generation to artifice. And the introduction of form and final causes in addition to material does not necessitate a craft structure.

Other Images and Metaphors

While Aristotle's references to artifice in *Generation of Animals* are heavily emphasised in the literature, Aristotle uses other images and metaphors that are less commonly acknowledged or only addressed through the craft analogy. This section considers these images on their own terms. Aristotle uses the images of the earth and celestial bodies, of rennet working on milk, and of a good householder.

The image of the earth and sun occurs early in *Generation of Animals* I. Aristotle uses it to explain the difference between the male and female with reference to the ways that people speak about the earth and sun:

> For by a male animal we mean that which generates in another (ἄρρεν μὲν γὰρ λέγομεν ζῷον τὸ εἰς ἄλλο γεννῶν), and by a female that which generates in itself (θῆλυ δὲ τὸ εἰς αὑτό); that is why in the macrocosm also, men think of the earth as female and a mother, but address heaven and the sun and other like entities as progenitors and fathers.[29]

The point is not so much that Aristotle views the cosmos itself as gendered, but that he draws on the way people talk about the earth and the sun as a way to explain the difference between male and female. The example depends on a generally accepted or at least acknowledged notion that the earth generates in itself and the sun generates in another. The precise senses that earth generates in itself and sun generates in another will clarify how the female generates in herself and the male in another. One view circulating among the Greeks is that the female is like the earth or the soil that is ploughed by the male. On this view, the female is just the place, but also on this view, the male is not

---

[29] *GA* 716a13–16.

contributing seed but rather the movement of the plough. The definitions of male and female that Aristotle offers here are not about what each contributes, which is not foreclosed by the definition, but neither is it the focus of it, but about where each generates. Earth generates in itself. It would be wrong to say that earth generates merely by being a place. Earth provides the soil, which is the nutriment as well as that which constitutes the seed. The sun generates by its vital heat. It is difficult to construe a way that the sun is imposing form externally when it generates in another. Surely the source of life is in the sun, but it is also in a sense in the earth. The sun makes the life potential in the seed in the earth together with water possible, drawing it forth from out of itself rather than imposing a shape on it. The metaphor to the earth and sun helps explain the difficulty Aristotle has in explaining how the semen can do the work of form without itself being the form of the human. The sun has to have a kind of power to affect the earth, which is vital heat, the kind of heat that animates. The sun is this vital heat, that is its form and actuality, and it is through this heat having a direct effect on the earth – the earth gets hot – that plants grow in the earth.

The second alternative to the artifice image Aristotle offers in *Generation of Animals* is fig-juice or rennet. Aristotle uses examples from artifice – of builders, carpenters and potters – in seven places in *Generation of Animals* in different contexts and working in various ways. He uses the rennet metaphor four times. Rennet then seems like a significant image for him. In this section, I consider the way that the rennet metaphor works to explain the work of semen and how it is different from the artifice metaphor.

The first passage in which Aristotle references fig-juice or rennet comes immediately before the first passage in *Generation of Animals* where Aristotle references the craft analogy. In this passage, recall, Aristotle is explaining that semen does not come from the whole of the body for this could not explain how the male and female contribute something different, which they do, the male contributing form and the female material. Aristotle explains the distinction thus:

> In fact, as in the coagulation of milk (γάλακτος πήξει), the milk being the material (τὸ μὲν σῶμα τὸ ὕλην), the fig-juice or rennet (ἡ πυτία) is that which contains the curdling principle (τὸ τὴν ἀρχὴν ἔχον τὴν συνιστᾶσαν), so acts the secretion of the male, being divided into the parts (μεριζόμενον) in the female.[30]

Aristotle describes the milk as the material and the work of the male contribution as akin to the work of fig-juice or rennet in the milk. The fig-juice or

---

[30] *GA* 729a11–14.

rennet is in the milk and works on the milk from being mixed within it, as an enzyme is a catalyst for a chemical reaction by being directly mixed with the chemicals whose reaction it causes. Acting in this way, the male contribution can cause a change in the female material from within it, from being in direct contact with it. In the next chapter, where Aristotle discusses the work of semen with reference to fig-juice, he explains that like the fig-juice, the male contribution can act upon the female contribution without becoming a material part of the result of generation:

> This material of the semen (τὸ σῶμα τῆς γονῆς) dissolves and evaporates because it has a liquid and watery nature. Therefore we ought not to expect it always to come out again from the female or to form any part of the embryo that has taken shape from it; the case resembles that of the fig-juice (τὸν ὀπὸν) which curdles milk (τὸν τὸ γάλα συνιστάντα), for this too changes without becoming any part of the curdling masses (τῶν συνισταμένων ὄγκων).[31]

The male contribution follows the metaphor of fig-juice by acting on the female contribution from within it, by curdling it. The fig-juice does not carry a form for the outcome. It acts by being the kind of material that it is in relation to the kind of material that the milk is. The result is not a mixture of milk plus fig-juice. The result is a transformation of the milk that the fig-juice effects. The result cannot be understood in any way as including fig-juice, and yet, it would not be what it is without the fig-juice. Aristotle can say in the next chapter, GA II.4, that while the female provides a body and a material mass, the male does not 'because it is not within what is produced that the tools or the maker must exist'.[32] The male contribution does work within the female contribution, but the material that makes it do its work does not become a part of what is formed. The rennet forms the female contribution, which then becomes the form on its own. In this way, the male forming the embryo differs from the rennet and milk example because the milk does not internalise a capacity to curdle further. Aristotle's point is that once the male principle does its work, the work is under way and it as an enmattered force does not remain in what is formed. This process becomes clearer in the last two passages where Aristotle references fig-juice or rennet.

> When the material secreted by the female in the uterus has been fixed (συστῇ) by the semen of the male (this acts in the same way as rennet acts upon milk (ἐπὶ τοῦ γάλακτος τῆς πυετίας), for rennet (ἡ πυετία) is a kind

---

[31] GA 737a10–15.
[32] GA 738b24–5.

of milk containing vital heat (γάλα ἐστὶ θερμότητα ζωτικὴν), which brings into one mass and fixes (συνίστησι) the similar material, and the relation of the semen to the menstrual blood in the same, milk and the menstrual blood being of the same nature) – when, I say, the more solid part comes together, the liquid is separated off from it, and as the earthy parts solidify membranes form all round it; this is both a necessary result and for the sake of something (καὶ ἐξ ἀνάγκης καὶ ἕνεκά τινος), the former because the surface of mass must solidify on heating as well as on cooling, the latter because the foetus must not be in a liquid but be separated from it.[33]

Like rennet, semen works by fixing or setting the menses. The rennet is able to compose, fix, condense – συνίστησι – the milk in a similar way that the semen is able to do this work in the menses – through vital heat. Aristotle says that rennet contains θερμότητα ζωτικὴν, the heat that brings life into animals. Milk and menses are of the same nature, Aristotle explains. Both are homogeneous parts that are characterised as condensed nutriment. Both are capable of being acted on in particular ways by heat to become something both similar to itself as the material and wholly other. Both become solid through heat having been fluid. Both are transformed while what they are remains at work in what is formed. While this material condenses and forms the cheese or the foetus, the liquid that is the rennet or the semen does not condense and form, it motivates the transformation. Aristotle once again addresses the details of how the semen works on the female contribution with reference to rennet *Generations of Animals* IV.4.

Whether the semen of the male contributes (συμβάλλεται) to the material of the embryo by itself becoming a part of it (μόριον γιγόμενον) and mixing with the semen of the female (τῷ τοῦ θήλεος σπέρματι μιγνύμενον), or whether, as we say, it does not act in this way but brings together and fashions (συνάγον καὶ δημιουργοῦν) the material within the female (τὴν ὕλην τὴν ἐν τῷ θήλει) and the generative secretion (τὸ περίττωμα τὸ σπερματικόν) as the fig-juice (ὁ ὀπὸς) does the liquid substance of milk (τὴν ὑγρότητα τοῦ γάλακτος), what is the reason why it does not form a single animal of considerable size? For certainly in the parallel case the fig-juice is not separated (κεχώρισται) if it has to curdle (συνιστάναι) a large quantity of milk, but the more (πλεῖον) the milk and the more (πλείων) the fig-juice put into it, so much the greater is the curdled mass . . . Now since it appears that the secretion of the female and that of the male need

[33] *GA* 739b20–9.

to stand in some proportionate relation to one another (I mean in animals of which the male emits semen), what happens in those that produce many young is this: from the very first the semen emitted by the male has power (δυνάμενον), being divided (μεριζόμενον), to form (συνιστάναι) several embryos, and the material contributed by the female is so much that several can be formed out of it. (The parallel of curdling milk (γάλακτος), which we spoke of before, is no longer in point here, for what is formed by the heat of the semen (τοῦ σπέρματος θερμότης) is not only of a certain quantity but also of a certain quality, whereas the fig-juice (ἐν τῷ ὀπῷ) and rennet (ἡ πυετία) quantity alone is concerned.) This then is just the reason why in such animals the embryos formed are numerous and do not all unite into one whole; it is because an embryo is not formed out of any quantity you please, but whether there is too much or too little, in either case there will be no result, for there is a limit set alike to the power of the heat which acts and to the material so acted upon (ὥρισται γὰρ ἡ δύναμις καὶ τοῦ πάσχοντος καὶ τῆς θερμότητος τῆς ποιούσης).[34]

This passage goes some distance to contrast the way semen works to the way that rennet works. Rennet requires a proper proportion in terms of quantity, while the relation of semen to menses must be proper in terms of quantity and quality. The context of the chapter is the cause of multiple births. Aristotle has earlier said that more semen is required in the generation of larger animals, but less in smaller animals, and that is why more small animals can be produced at once.[35] In the case of fig-juice working on milk, the more milk and the more fig-juice there is, the larger is the product that is curdled. What is required is the proper relation. Having more milk as long as there is also more rennet or fig-juice could continue to extend without preventing curdling. The proportional ratio that determines the curdling effect stands in contrast to the way that material can overwhelm the semen – if there is too much fat, for example, it can fail to properly concoct the menses.[36] In animals where there are multiple births, Aristotle says that the male has the power to form several embryos if it is divided. The semen works by dividing the material because there is a sufficient amount of it as well as a sufficient amount of menses. Multiple offspring are produced from the proper proportion of heat to the material – so when there is more heat and more material in the proper proportion, there can be more offspring produced. The insufficiency of the

[34] *GA* 771b18-28, 772a16-29.
[35] *GA* 771a31-3.
[36] *GA* 727a34-727b1.

metaphor to rennet is not a difference between how the rennet works and how the semen works, but rather a difference in what it works on. As Aristotle notes, while the quality of what is worked on is not relevant for the rennet, it is relevant for the semen. An embryo is formed out of both the proper quality of menses and the proper quantity.

The last image to consider is of nature as a good householder. In *Generation of Animals* II.6, Aristotle is explaining the order of the generation of parts, the internal coming to be before the external parts. In this context, Aristotle argues that air differentiates the parts. We know that air is present because heat, which acts, and moisture, which is acted upon, are present. Aristotle then distinguishes between the end and that which exists for the end, and within things which exist for the end, that which is the origin of movement and that which is used by the end. For example, Aristotle distinguishes between that which can generate and that which is a tool for what is generated. That which generates exists prior to the tool for what is generated. Aristotle divides the relation to the end into three: the end, the principle of movement and generation which exists for the sake of the end, and what the end uses. The first things that comes into being in an animal contain the principle and end of their nature – this principle is considered both as the moving power and with the whole natural thing as part of the end. But as Aristotle explains, 'So it is not easy to distinguish which of the parts are prior, those which are for the sake of another or that for the sake of which are the former.'[37] Semen would seem to be a good example of this difficulty, since on the one hand, it is for the sake of the end, which is the natural living organism, and on the other hand, it itself contains the end or soul. Semen seems to be the mover that also contains the essence, yet semen is also a mover that pre-exists the formed essence in the embryo. Aristotle contrasts changing things to unchanging things by saying that the first principle of a changing thing is the principle of movement, distinguished from the essence, which in blooded things is the heart.

In what follows, Aristotle explains how change is motivated in natural things through the work of heat and cold on the moist and dry. The homogeneous parts are formed from heat and cold, and the differences between them are based on the degree of heat or presence or absence of moisture. The sinews and bones, Aristotle explains, are formed from the internal heat out of that which is 'naturally fitted' for the purpose. At this point, Aristotle turns to the way that semen works through heat and movement, recalling that in natural things, the moving cause is prior. The male parent imparts movement through

---

[37] *GA* 742b7–8; see 742a19–742b8.

heat. Aristotle explains that nature works through cold and heat, which work to bring about different results in the forming of the embryo. What follows is a detailed account of how heat and cold work to form flesh, skin, fat and the heart. The heart, Aristotle explains yet again, is the first part of the animal as a whole, which enables sensation. The heat of the heart is the cause of the formation of other organs, including the brain, which is formed through cold (because all the heat is in the heart).

After an account of how different sense organs are formed after the first principle, Aristotle makes a more general claim about how nature works in light of this explication of the development of parts in the embryo: 'But since nature makes nothing superfluous or in vain, it is clear also that she makes nothing too late or too soon, for if she did the result would be either in vain or superfluous.'[38] As an example, Aristotle offers a material explanation of why eyes are formed when they are: the proper degree of concoction has been achieved in the brain at this point. The explanation is not simply material: those parts that are closer to the ruling principle of the natural thing are formed out of the 'nutriment that is first and purest and fully concocted', while those parts needed for these parts are formed from 'inferior nutriment' and 'residues left over' from the other parts. Aristotle explains why this is the case:

For nature, like a good householder (οἰκονόμος ἀγαθός), is not in the habit of throwing away anything from which it is possible to make anything useful. Now in a household (ἐν δὲ ταῖς οἰκονομίαις) the best part of the food that comes in is set apart for the free men, the inferior and the residue of the best for the slaves, and the worst is given to the animals that live with them. Just as the intellect acts thus from outside (ὁ θύραθεν ταῦτα ποιεῖ νοῦς) with a view to the growth of the persons concerned, so in the case of the embryo itself does nature form (συνίστησιν) from the purest material (ὕλης) the flesh and the body of the other sense-organs, and from the residues thereof bones, sinews, hair, and also nails and hoofs and the like; hence these are last to assume their form (σύστασιν), for they have to wait till the time when nature has some residue to spare (περίττωμα).[39]

In comparing nature to a householder, Aristotle makes nature into a question of proper distribution, a question of which resources should go to what parts. The best resources, following the Athenian household model, go to the best, which for Aristotle is those who direct and organise the household,

[38] *GA* 744a36-744b1.
[39] *GA* 744b15-26.

while the inferior resources go to those who put to work the orders of the free men – the slaves. The worst resources go to those who do not in any way deliberate or respond to deliberative commands – the animals. In the *Politics*, Aristotle makes an argument for natural slavery that sets the terms so extremely that no one would seem to qualify, an implicit argument against the conventional slavery practices of his time, while also acknowledging that slaves respond to reason and can improve their reasoning.[40] His argument in support of defining the human being as having λόγος by contrast to animals who have only voice treads a very thin line that requires determination of whether animal responses to pain and pleasure signal just that or benefit and harm as well. That thin line suggests that the work of drawing this well-structured household is more of a political dispute than technocratically and hierarchically determinable.

Nature acts like a householder in Aristotle by organising resources toward what is best in the body. Political life is a matter of an engaged dispute requiring deliberation over what is best and how to organise the community. In the *Politics*, Aristotle maintains that deliberation over the beneficial and harmful, the just and the unjust is what forms the communities of the household and the polis. In both, we dispute what is best and how to manifest what is best, as the body generally organises resources in a way that serves the central function of living. This discussion of nature as a good householder follows Aristotle's argument that the household is hierarchically ordered in contrast to the equality and freedom of the polis, while also echoing the ways that the household is also a place for deliberation and judgement. Nature is a good householder because nature is a good steward of resources – using everything that can be used. Like a householder, nature arranges the nutriment in the natural living body so that it uses the best material for the best parts. Aristotle draws a parallel between the way that the intellect acts and the way that nature acts in forming the embryo. Nature acts from the inside the way that the intellect acts from the outside. Here the comparison is not of nature to craft in the sense of a builder, that is, to the intellect in the mind of the craftsperson. If there is any τέχνη, it would be more like the medical art, which aims toward the growth of the person by considering what constitutes health and how to motivate the body to achieve it. The intellect is external insofar as it is an external source of order that achieves growth by contrast to nature which is an internal source of order that draws on material at different levels of concoction or formation for different parts.

---

[40] *Pol.* 1254b16–24, 1259b28–9, 1260b5–7, 1253b34. See Adriel M. Trott, *Aristotle on the Nature of Community*, pp. 178–92, and Frank, *A Democracy of Distinction*, pp. 17–53.

Aristotle uses the word συνίστησιν to describe what nature does to make flesh and body out of material. This word is translated as form, but it is not so much a technical sense of form as it is the particular kind of work that occurs on this material, composing or compressing or constituting. What then is formed – σύστασιν – is what has been so compressed to form particular parts. Nature is not working by moving material into an organisation from outside of it with no contact with it, but by pressing upon it from within it. As a householder, nature acts on the purest material to form flesh and sense organs and then on what is left over to form the other parts. In the order of development, menses is the initial best material, which is animated fully by the semen at the moment the best part, the heart – the source of blood – is formed. The stomach, which processes and develops nutriments for the sake of growth and development of the nutriment, follows soon after. In the processes of forming, the excess material from each process becomes the material for the next parts to be formed. The menses that becomes the material for these central parts is itself formed as semen is from the last useful left over parts, the useful residue. The good householder makes the semen as the source of animated life and the menses from the same last part, the useful residue. The householder image shows how nature works from the material available to it to differentiate the very parts that are considered the opposing poles of a gender binary and metaphysically distinct causes.

Conclusion: On Material in Aristotle's Biology

Readers of Aristotle have long construed form and matter as contraries. Form is taken to be the source of order and definition; material is what must be ordered and defined. Material is what lacks form. In his biology, Aristotle associates form with male and matter with female at the head of an extended tradition that counts male as properly human and capable of giving definition to the world and female as what is in need of proper boundaries, lacking meaning and significance of her own. This strict division between form and matter makes them both absolutely other and defined as an inversion or a lack of the other with form on the positive side. Such a strict division supports a conception of nature that has form shaping and arranging unshaped stuff. Yet Aristotle's account of generation in animals suggests that the strict division between form and matter is untenable. He describes the work of that which brings forth soul in the embryo in material terms. He describes the work in terms of the material power of heat made internal. He explains how material has certain capacities that enable it to become the embryo. The analogies that Aristotle uses to explain how semen can come from the father but be the cause of soul in the offspring, for example, lead commentators to emphasise the sense in which semen is a tool of form, and so not form itself. Such readings preserve the notion that form is intelligible and operating at a distance from material, even though material is what is unified and constituted by soul or form. This book has worked to tease out these ways that form is working in and through matter and that matter has a character of its own.

In conclusion, I consider some further thoughts for how the conception of form and matter on something like a Möbius strip model might be useful for understanding Aristotle and for resisting a metaphysics that by opposing form to matter subordinates material's work to form's. Aristotle introduces the four causes in response to the materialist arguments that material is sufficient to explain what is. The materialists can explain things either according to necessity or contingency, but not according to a regular process of change. According to necessity, what is exists because material necessarily acts the way that it does. According to contingency, the changes between things occur because certain materials join and disjoin from other materials, condense and rarify to form new materials without a determinable reason.

Material is not without causal power for Aristotle. But material alone is. Material by itself can explain change either as haphazard and irregular or as necessarily occurring in the same way according to the nature of the matter. Matter by itself cannot offer reasons for the regularity of the diversity of natural substances, but matter helps explain how and why they are generated and destroyed. The reasons for form show that Aristotle does not need to make form immaterial for it to do its work. He just needs to argue that form is not reducible to material, which does not mean that it cannot work in, through and with material. Form's power is distinct from material's power, but it is dependent in natural generation on this power for it to be the power that is formal.

This dependence is at work in artifice as it is in nature, but in artifice the separability of the roles of form and matter seem more evident. The artisan needs material not only to work on and shape, but both to do the shaping in the form of tools and to shape and form the tools. The artisan is considered the source of the form and the tools are the material mediators whereby the artisan manifests the form in the artefact. In the artificial thing, the form can be at once in the head of the artisan and in the artificial thing because the form is a pattern or a shape. But natural things are crucially different from artificial things for Aristotle. The form is not mere shape in a natural thing but an organising cause. As an organising cause, it actively organises that of which it is the form, not as a tool might do in order to set the material in place, but persistently, in order to maintain itself as what it is. Semen is the source of this organising principle from the father. It takes over from the father the internal heat that makes nutritive soul capable of bringing forth life. If semen is a tool, it is indistinguishable from the user of the tool, which is the form or soul. Soul is what it is for a natural substance to be. As this active potential for life, soul works through semen. It is tempting to think of semen as the pineal gland of the ancients, which is how many commentators seem to discuss it. On that view, it seems to be a mediator between mind and body, separate entities that need to be joined. But soul is not an entity in this sense for Aristotle. It is the actuality of body, the source of life in the body. As this source, semen both is soul and the tool of the soul: it is the power of soul in its material composition. Material cannot do the organising work itself, but material can enable form to do the organising work in material. And in that sense, natural substances are working on themselves to fulfil their end.

It is also because form is an organising cause that the form in the father cannot be the same as the form in the offspring in the way that the form in the mind of the artist can be the same as that in the artefact. The form in natural substances is not a pattern or an arrangement of material. As an organising cause, form must be particular to each natural substance. And the form of the offspring becomes particular to it by being internalised in the semen, no longer depending on the father for its heat and source of life, as blood is. The heat is the source of power of the semen, but not just as heat. Semen is heat as the internalised power, as nature is the internal source of movement. Semen both is and is not yet living itself. It has an internalised source of movement, but it is the source of the life that will become, with the menses, a living continuously self-organising natural substance.

Aristotle calls reproduction 'the most natural act', 'the goal toward which all things strive, that for the sake of which they do whatsoever their nature renders possible'.[1] Semen might seem peculiar in having a principle that is both for itself and the capacity for life in another, but this self-relation that is also for another is the fulfilment of the first substance. Natural things in Aristotle aim to fulfil themselves by, in a sense, becoming another. I maintain that the actuality in generation is akin to the actuality in persistent substance, not because natural substance is fundamentally of the structure of artefacts, but because form and matter are interwoven as much in natural generation as in hylomorphic natural substance.

Revisiting the One-Sex and Two-Sex Models

The one-sex and two-sex models present different problems for how a nonreductive difference can obtain between male and female, and further between form and material. The one-sex model establishes male as the measure of what is worthy and meaningful and the female as what is measured by the male and considered in proximity from the male. The two-sex model acknowledges essential difference between male and female and results in a hierarchy based on valuing the essence of male as what can overcome the body by contrast to the female who is defined by and restricted to her reproductive capacity. The two models of sexual difference describe well the problems for thinking true difference between form and matter and the gender implications for those problems. Luce Irigaray's critique of hylomorphic metaphysics in Aristotle is that form is considered prior in being and in knowledge, capable of being independent from material, while material is dependent on form for its being

---

[1] *De An.* 415a27–415b1.

and for being known. And yet, Irigaray argues, form's power depends on bringing material under its dominion. In Aristotle, natural form depends on material to generate new substance. Irigaray aims to revive material's power by showing how it is the foundation for form's power. Judith Butler raises the concern that such a view would seem to naturalise material and further affirm the characterlessness of matter ignoring the way that the natural givenness of material is itself constructed through this binary.

The troubled conception of the difference between material and form is traced to Aristotle's metaphysics and biological works. This book offers an alternative to that view, one that shows material's independence in Aristotle's metaphysics and natural form's multiple ways of being dependent on material. The argument does not simply invert the structure and thereby produce the same problem wherein form's priority in definition and in generation are maintained because of work material does for it. Material comes to appear under its own power and not from the recognisability it only draws from form.

Form is multiply related to material in natural substances and natural generation. Material works in semen. Semen is distinguished from blood by its capacity to internalise the heat for which blood remains dependent on some external force. Form requires available material in the moment of generation in menses, which is distinguished from semen by different degrees of concoction. In natural substances, the functional matter is organised by form to achieve the natural substance's end.

Material has power of its own at the level of elemental forces and the elemental. Cold, the element associated with female, proves to be an active and positive power. Moisture, too, considered a passive elemental force, has its own character that contributes to the power of vital heat in semen, as well as making menses the kind of material capable of becoming alive. Semen and menses develop out of the same material and become differentiated through a heat that characterises both but differentiates them through degrees. Form acts on material in generation as like on like, as having the same material basis that enables form to actualise what is already potential in the material. Semen seems like a luxury part because it is formed from residue, and yet semen does the work of a primary part, perhaps the most primary part, the form or soul.

Considered in terms of the one-sex and two-sex models, Aristotle's account of form and matter in generation shows how the strict distinction between form and matter becomes a difference on a continuum, the formal difference becomes a material difference. The difference between form and matter appears almost impossible to think because our ways of thinking difference are either formal or material. To think the difference between these differences appears impossible. If it is formal, form is the measure of the difference between them, which does not tell us so much what material is, just that it is not form, and hence wholly otherwise than form. If it is material, matter is the measure of the difference, and form is distinguished from matter through material power, which seems to make form material, and so not wholly otherwise than material. This material difference challenges the hierarchy between form and matter by showing form's dependence on material, but it raises the question of whether form and matter are different powers or reducible to material. This comparison between formal difference and material difference points to the difficulty in thinking their difference that can be found in Aristotle's account of the difference between form and matter in generation. This difference can be captured in the image of the Möbius strip.

The Möbius Strip

A Möbius strip is a one-sided surface formed out of a two-sided plane to produce a three-dimensional shape. It allows us to think the relation of difference that is interrelated without being reducible of one to the other where what is outside becomes inside and outside again. It does not deny the difference between the sides, which can be articulated at particular points, but it allows those differences to move into and affect one another without being reducible to the other. Form and matter have traditionally been understood in Aristotle as opposite sides of the strip, though as the one- and two-sex models show, both difference as opposition and difference as lack models set up a hierarchy between form and matter. The Möbius strip produces a model whereby a torsion between form and matter generates substance.

In her use of the Möbius strip model, Elizabeth Grosz argues that a limitation of the model is that it is 'not well suited for representing modes of becoming, modes of transformation'.[2] The argument this book presents for the conjunction of form and matter in generation, where form works through material and material has its own character, shows that the Möbius strip is also useful for thinking generation and transformation. Notably, mathematicians and physicists have shown that Möbius strip shapes that occur in the world – in crystals and lightwaves, for example – go beyond an inelastic or indevelopable plane.[3] The torsion increases the energy beyond what exists on both sides prior to the turn that creates the strip.

---

[2] Elizabeth Grosz, *Volatile Bodies: Toward a Corporeal Feminism*, p. 210.
[3] E. L. Starostin and G. H. M. van der Heijden, 'The Shape of a Möbius Strip', *Nature Materials* 6 (2007), pp. 563–7; S. Tanda, T, Tsuneta, Y. Okajima, K. Inagaki, K. Yamaya and N. Hatakenaka, 'A Möbius Strip of Single Crystals', *Nature* 417 (23 May 2002), 397–8; 'Light Twists Like a Möbius Strip', *New Scientist* 225 (7 February 2015).

This book makes the case that Aristotle's hylomorphism as articulated in the biological works and explained in the theoretical works points to this kind of interdependent relationship between form and matter in natural substance. In doing so, it defends the view that Aristotle's conception of nature is not reducible to a craft model where form is imposed on material. This reading does not undo Aristotle's association of the male with form, which Aristotle credits with being more fully definitive and essential to what natural substance is. But it does reject the view that natural form is capable of a kind of independence that would permit it to dominate material. Its own dependence on material to fulfil its work points to the power of material and to the extent to which form can only be definitive and essential to the extent that material enables it to be. Aristotle remains committed to a hylomorphism that seems to go all the way down in natural substance so that even form in natural substance works through material and even material is already of a certain character and power. What was presumed independent in form is more dependent and what was presumed wholly dependent in material is independent. This account goes some way towards undoing the maintenance of the gender binary around the distinction between what has the capacity to form and affect the world and what needs to be formed and affected.

Bibliography

Primary Ancient Sources

Aeschylus, *Seven against Thebes*, trans. Anthony Hecht and Helen H. Bacon (London: Oxford University Press, 1974).

Albertus Magnus, *Metaphysica 16/1*, in Thomas Marschler (ed.), *Editio Coloniensis* (Münster: Aschendorff, 2015).

Albertus Magnus, *On Animals*, ed. Irven Michael Resnick and Kenneth Kitchell (Baltimore: Johns Hopkins University Press, 1999).

Albertus Magnus, *Physica 4/1*, in Thomas Marschler (ed.), *Editio Coloniensis* (Münster: Aschendorff, 2015).

Alexander of Aphrodisias, *Alexander of Aphrodisias: On Aristotle's* Meteorology 4, trans. Eric Lewis (Ithaca, NY: Cornell University Press, 1996).

Apollodorus, *The Library, Volume I: Books 1–3.9*, trans. James G. Frazer (Cambridge, MA: Harvard University Press, 1921).

Aristotle, *Aristotle: De Anima*, trans. Mark Shiffman (Newburyport, MA: Focus Publishing, 2011).

Aristotle, *Aristotle: Metaphysics Books Z and H*, trans. David Bostock (Oxford: Clarendon Press, 1994).

Aristotle, *Aristotle on Coming-To-Be and Passing-Away (De Generatione et Corruptione): A Revised Text with Introduction and Commentary*, trans. Harold H. Joachim (Oxford: Clarendon Press, 1922).

Aristotle, *Aristotle: Physics Books I and II*, trans. William Charlton (Oxford: Clarendon Press, 1992).

Aristotle, *Aristotle's* De Generatione et Corruptione, trans. C. J. F. Williams (Oxford: Clarendon Press, 1982).

Aristotle, *Aristotle's* De Partibus Animalium I *and* De Generation Animalium I, trans. D. M. Balme (Oxford: Clarendon Press, 1972).

Aristotle, *The Complete Works of Aristotle: The Revised Oxford Translation*, ed. Jonathan Barnes, 2 vols (Princeton: Princeton University Press, 1984).

Aristotle, *De Caelo*, trans. W. K. C. Guthrie (Cambridge, MA: Harvard University Press, 1939).

Aristotle, *Generation of Animals*, trans. A. L. Peck (Cambridge, MA: Harvard University Press, 1942).

Aristotle, *History of Animals Books 1–3*, trans. A. L. Peck (Cambridge, MA: Harvard University Press, 1965).

Aristotle, *Meteorologica*, trans. H. D. P. Lee (Cambridge, MA: Harvard University Press, 1952).

Aristotle, *On Sophistical Refutations, On Coming-to-Be and Passing Away, On the Cosmos*, trans. E. S. Forster and D. J. Furley (Cambridge, MA: Harvard University Press, 1955).

Aristotle, *On the Soul, Parva Naturalia, On Breath*, trans. W. S. Hett (Cambridge, MA: Harvard University Press, 1957).

Aristotle, *Parts of Animals, Movement of Animals, Progression of Animals*, trans. A. L. Peck and E. S. Forster (Cambridge, MA: Harvard University Press, 1961).

Craik, Elizabeth M., *The 'Hippocratic' Corpus: Content and Context* (New York: Routledge, 2015).

Curd, Patricia (ed.), *A Presocratics Reader: Selected Fragments and Testimonia*, trans. Richard McKirahan and Patricia Curd (Indianapolis, IN: Hackett, 2011).

Galen, *On the Usefulness of the Parts of the Body*, trans. Margaret Tallmadge May (Ithaca, NY: Cornell University Press, 1968).

Hesiod, *Theogony, Works and Days, Shield*, trans. Apostolos N. Athanassakis (Baltimore: Johns Hopkins University Press, 2004).

Hippocrates, *Affections, Diseases I and II*, trans. Paul Potter (Cambridge, MA; Harvard University Press, 1988).

Hippocrates, *Ancient Medicine, Air, Water, Places, Epidemics I and III, The Oath, Precepts* and *Nutriment*, trans. W. H. S. Jones (Cambridge, MA: Harvard University Press, 1923).

Hippocrates, *Generation, Nature of the Child, Diseases IV, Nature of Women* and *Barrennes*, trans. Paul Potter (Cambridge, MA: Harvard University Press, 2012).

Hippocrates, *Nature of Man, Regimen in Health, Humours, Aphorisms, Regimen I, II and III* and *Dreams*, trans. W. H. S. Jones (Cambridge, MA: Harvard University Press, 1931).

Hippocrates, *Places in Man, Glands, Fleshes, Prorrhetic I and II, Physician, Use of Liquids, Ulcers, Haemorrhoids* and *Fistulas*, trans. Paul Potter (Cambridge, MA: Harvard University Press, 1995).

Homer, *Homeric Hymn to Demeter: Translation, Commentary and Interpretative Essays*, ed. Helene P. Foley (Princeton: Princeton University Press, 1994).

Homer, *The Homeric Hymns*, trans. Apostolos N. Athanassakis (Baltimore: Johns Hopkins University Press, 2004).

Kirk, G. S., J. E. Raven and Malcolm Schofield, *The Presocratic Philosophers: A Critical History with a Selection of Texts* (Cambridge: Cambridge University Press, 1984).

Plutarch, *Moralia, Volume III*, trans. Frank Cole Babbitt (Cambridge, MA: Harvard University Press, 1931).

Proclus, *Proclus' Commentary on Plato's* Parmenides, trans. G. R. Morrow and J. M. Dillon (Princeton: Princeton University Press, 1987).

Sophocles, *Aeschylus I: Oresteia*, trans. Richard Lattimore (Chicago: University of Chicago Press, 1953).

Sophocles, *Sophocles I*, ed. David Grene and Richard Lattimore, trans. David Grene (Chicago: University of Chicago Press, 1991).

Literature

Adelman, Janet, 'Making Defect Perfection: Shakespeare and the One-Sex Model', in Viviana Comensolli and Anne Russell (eds), *Enacting Gender on the Renaissance Stage* (Urbana, IL: University of Illinois Press, 1998), pp. 23–52.

Adorno, Theodor W., *Metaphysics: Concepts and Problems*, trans. Edmund Jephcott (Stanford: Stanford University Press, 2001).

Albritton, Rogers, 'Forms of Particular Substances in Aristotle's *Metaphysics*', *Journal of Philosophy* 54.22 (1957), pp. 699–708.

Allen, Prudence, *The Concept of Woman: The Aristotelian Revolution 750 BCE–1250 AD* (Montreal: Eden Press, 1985).

Annas, Julia, 'Aristotle on Substance, Accident and Plato's Forms', *Phronesis* 22.2 (1977), pp. 146–60.

Balme, D. M., 'Aristotle's Biology Was Not Essentialist', *Archiv für Geschichte der Philosophie* 62.1 (1980), pp. 1–12.

Balme, D. M., 'The Place of Biology in Aristotle's Philosophy', in Allan Gotthelf and James G. Lennox (eds), *Philosophical Issues in Aristotle's Biology* (Cambridge: Cambridge University Press, 1987), pp. 9–20.

Balme, D. M., ''Άνθρωπος ἄνθρωπον γέννα: Human is Generated in Human', in Gordan R. Dunstan (ed.), *The Human Embryo: Aristotle and the Arabic and European Traditions* (Exeter: University of Exeter Press, 1990), pp. 20–31.

Baracchi, Claudia, *Aristotle's Ethics as First Philosophy* (Cambridge: Cambridge University Press, 2008).
Barnes, Jonathan, *Aristotle: A Very Short Introduction* (Oxford: Oxford University Press, 2001).
Bartoš, Hynek, 'Aristotle and the Hippocratic *De Victu* on Innate Heat and Kindled Soul', *Ancient Philosophy* 34.2 (2014), pp. 289-315.
Bianchi, Emanuela, *The Feminine Symptom: Aleatory Matter in the Aristotelian Cosmos* (New York: Fordham University Press, 2014).
Bleier, Ruth, *Science and Gender: A Critique of Biology and Its Theories of Women* (New York: Pergamon, 1984).
Blok, Josine, 'Sexual Asymmetry: A Historiographical Essay', in Josine Blok and Peter Mason (eds), *Sexual Asymmetry: Studies in Ancient Society* (Amsterdam: J. C. Gieben, 1987).
Blundell, Sue, *Women in Ancient Greece* (Cambridge, MA: Harvard University Press, 1995).
Bogen, James, 'Change and Contrariety in Aristotle', *Phronesis* 37.1 (1992), pp. 2-21.
Bos, Abraham P., *Aristotle on God's Life-Generating Power and on Pneuma as Its Vehicle* (Albany: State University of New York Press, 2018).
Boys-Stones, George, 'Physiognomy and Ancient Psychological Theory', in Simon Swain (ed.), *Seeing the Face, Seeing the Soul: Polemon's Physiognomy from Classical Antiquity to Medieval Islam* (Oxford: Oxford University Press, 2007).
Butler, Judith, *Bodies That Matter* (London: Routledge, 1993).
Byrne, Christopher, 'Matter and Aristotle's Material Cause', *Canadian Journal of Philosophy* 31.1 (2001), pp. 85-112.
Campese, Silvia, '*Donna, casa, città nell'antropologia di Aristotele*', in Silvia Campese, Paola Manuli and Giulia Sissa (eds), *Madre Materia: Sociologia e Biologia Della Donna Greca* (Turin: Boringhieri, 1983), pp. 147-92.
Campese, Silvia, Paola Manuli and Giulia Sissa (eds), *Madre Materia: Sociologia e Biologia Della Donna Greca* (Turin: Boringhieri, 1983).
Carson, Anne, 'Putting Her in Her Place', in David M. Halperin, John J. Winkler and Froma I. Zeitlin (eds), *Before Sexuality: The Construction of Erotic Experiences in the Ancient Greek World* (Princeton: Princeton University Press, 1990), pp. 135-69.
Cerami, Cristina, 'Function and Instrument: Toward a New Criterion of the Scale of Being in Arisotle's *Generation of Animals*', in Andrea Falcon and David Lefebvre (eds), *Aristotle's Generation of Animals: A Critical Guide* (Cambridge: Cambridge University Press, 2018), pp. 130-50.
Charlton, William, 'Prime Matter: A Rejoinder', *Phronesis* 28.2 (1983), pp. 197-211.

Code, Alan, 'An Aporematic Approach to Primary Being in *Metaphysics* Z', *Canadian Journal of Philosophy* 14 (sup1) (1984), pp. 1–20.

Code, Alan, 'Aristotle: Essence and Accident', in R. E. Grandy and R. Warner (eds), *Philosophical Grounds of Rationality: Intuitions, Categories, Ends* (Oxford: Oxford University Press, 1986), pp. 411–39.

Code, Alan, 'On the Origins of Some Aristotelian Theses about Predication', in J. Bogen and J. E. McGuire (eds), *How Things Are: Studies in Predication and the History and Philosophy of Science* (Dordrecht: Springer, 1985), pp. 101–31.

Code, Alan, 'The Persistence of Aristotelian Matter', *Philosophical Studies: An International Journal for Philosophy in the Analytic Tradition* 29.6 (1976), pp. 357–67.

Cohen, Sheldon, 'Aristotle on Heat, Cold, and Teleological Explanation', *Ancient Philosophy* 9.2 (1989), pp. 255–70.

Cohen, Sheldon, 'Aristotle's Doctrine of the Material Substrate', *The Philosophical Review* 93.2 (1984), pp. 171–94.

Cole, Eva Browning, 'Review of *Sowing the Body: Psychoanalysis and Ancient Representations of Women*, by Page DuBois', *American Philosophical Association Newsletter on Philosophy and Feminism* 89.1 (1989), pp. 88–9.

Colebrook, Claire, 'Dynamic Potentiality: The Body that Stands Alone', in Athena Athanasiou and Elena Tzelepis (eds), *Rewriting Difference: Luce Irigaray and 'the Greeks'* (Albany: State University of New York Press, 2010), pp. 177–90.

Coles, Andrew, 'Biomedical Models of Reproduction in the Fifth Century BC and Aristotle's *Generation of Animals*', *Phronesis* 40.1 (1995), pp. 48–88.

Connell, Sophia M., *Aristotle on Female Animals* (Cambridge: Cambridge University Press, 2016).

Cooper, John M., 'Aristotle on Natural Teleology', in Malcolm Schofield and Martha Craven Nussbaum (eds), *Language and Logos: Studies in Ancient Greek Philosophy Presented to G. E. L. Owen* (Cambridge: Cambridge University Press, 2006), pp. 197–222.

Cooper, John M., 'Metaphysics in Aristotle's Embryology', *Proceedings of the Cambridge Philological Society* 34 (1988), pp. 15–41.

D'Hoine, Pieter, 'Aristotle's Criticisms of Non-Substance Forms and Its Interpretation by the Neoplatonic Commentators', *Phronesis* 56.3 (2011), pp. 262–307.

Dancy, Russell, 'On Some of Aristotle's Second Thoughts about Substance: Matter', *Philosophical Review* 87.3 (1978), pp. 372–413.

Dean-Jones, Lesley, *Women's Bodies in Classical Greek Science* (Oxford: Clarendon Press, 1994).

Deslauriers, Marguerite, 'Sex and Essence in Aristotle's *Metaphysics* and Biology', in Cynthia Freeland (ed.), *Feminist Interpretations of Aristotle* (University Park: Pennsylvania State University Press, 1998), pp. 138–67.

DuBois, Page, *Sowing the Body: Psychoanalysis and Ancient Representations of Women* (Chicago: University of Chicago Press, 1988).

Elshtain, Jean Bethke, *Public Man, Private Woman: Women in Social and Political Thought*, 2nd edn (Princeton: Princeton University Press, 1993).

Falcon, Andrea, and David Lefebvre (eds), *Aristotle's Generation of Animals: A Critical Guide* (Cambridge: Cambridge University Press, 2018).

Fielding, Helen, 'Questioning Nature: Irigaray, Heidegger and the Potentiality of Matter', *Continental Philosophy Review* 36.1 (2003), pp. 1–26.

Finley, M. I., *Democracy Ancient and Modern* (New Brunswick, NJ: Rutgers University Press, 1973).

Flemming, Rebecca, *Medicine and the Making of Roman Women: Gender, Nature, and Authority from Celsus to Galen* (Oxford: Oxford University Press, 2000).

Frank, Jill, *A Democracy of Distinction: Aristotle and the Work of Politics* (Chicago: University of Chicago Press, 2005).

Frede, Michael, 'Substance in Aristotle's Metaphysics', in Allan Gotthelf (ed.), *Aristotle on Nature and Living Things: Philosophical and Historical Studies Presented to David M. Balme on His 70th Birthday* (Pittsburgh: Mathesis Publications, 1985), pp. 17–26.

Frede, Michael, and Gunther Patzig, '*Sind Formen Allgemein oder Individuell?*' in Michael Frede and Gunther Patzig (eds and trans.), *Aristoteles Metaphysics Z: Text, Übersetzung, und Kommentar*, vol. 1 (Munich: Verlag C. H. Beck, 1988), pp. 48–57.

Freeland, Cynthia A., 'Nourishing Speculation: A Feminist Reading of Aristotelian Science', in Bat-Ami Bar On (ed.), *Engendering Origins: Critical Feminist Readings in Plato and Aristotle* (Albany: State University of New York Press, 1994), pp. 145–88.

Freudenthal, Gad, *Aristotle's Theory of Material Substance* (Oxford: Clarendon Press, 1999).

Furth, Montgomery, *Substance, Form and Psyche: An Aristotelian Metaphysics* (Cambridge: Cambridge University Press, 1988).

Furth, Montgomery, 'Transtemporal Stability in Aristotelian Substances', *Journal of Philosophy* 75.11 (1978), pp. 624–46.

Gelber, Jessica, 'Females in Aristotle's Embryology', in Andrea Falcon and David Lefebvre (eds), *Aristotle's Generation of Animals: A Critical Guide* (Cambridge: Cambridge University Press, 2018), pp. 171–87.

Gelber, Jessica, 'Form and Inheritance in Aristotle's Embryology', *Oxford Studies in Ancient Philosophy* 39 (2010), pp. 183–212.

Gill, Mary Louise, *Aristotle on Substance: The Paradox of Unity* (Princeton: Princeton University Press, 1989).
Gill, Mary Louise, 'Aristotle's *Metaphysics* Reconsidered', *Journal of the History of Philosophy* 43.3 (2005), pp. 223–51.
Gill, Mary Louise, 'The Limits of Teleology in Aristotle's *Meteorology* IV.12', *HOPOS: The Journal of the International Society for the History of Philosophy of Science* 4.2 (2014), pp. 335–50.
Gomme, A. W., 'The Position of Women in Athens in the Fifth and Fourth Centuries', *Classical Philology* 20.1 (1925), pp. 1–25.
Gotthelf, Allan, 'Aristotle's Conception of Final Causality', in Allan Gotthelf and James G. Lennox (eds), *Philosophical Issues in Aristotle's Biology* (Cambridge: Cambridge University Press, 1987), pp. 204–42.
Gotthelf, Allan, 'Notes toward a Study of Substance and Essence in Aristotle's Parts of Animals II–IV', in *Teleology, First Principles and Scientific Method* (Oxford: Oxford University Press, 2011), pp. 217–40.
Green, Judith M., 'Aristotle on Necessary Verticality, Body Heat, and Gendered Proper Places in the Polis: A Feminist Critique', *Hypatia* 7.1 (1992), pp. 70–96.
Grosz, Elizabeth, *Volatile Bodies* (Bloomington: Indiana University Press, 1994).
Hansen, M. H., *Democracy in the Age of Demosthenes* (Oxford: Blackwell Publishers, 1991).
Hanson, Ann Ellis, 'Conception, Gestation, and the Origin of Female Nature in the *Corpus Hippocraticum*', *Helios* 19.1–2 (1992), pp. 31–71.
Haraway, Donna, 'Situated Knowledges: The Science Question in Feminism and the Privilege of Partial Perspective', *Feminist Studies* 14.3 (1988), pp. 575–99.
Heidegger, Martin, *Basic Concepts of Aristotelian Philosophy*, trans. Robert D. Metcalf and Mark B. Tanzer (Indianapolis: Indiana University Press, 2009).
Heidegger, Martin, 'On the Essence and Concept of Φύσις in Aristotle's *Physics* B, 1', in William McNeill (ed.), *Pathmarks* (Cambridge: Cambridge University Press, 1998).
Heidegger, Martin, *Plato's* Sophist, trans. Richard Rojcewicsz and André Schuwer (Indianapolis: Indiana University Press, 1997).
Henry, Devin M., 'Aristotle on the Mechanism of Inheritance', *Journal for the History of Biology* 39.3 (2006), pp. 425–55.
Henry, Devin M., 'How Sexist is Aristotle's Developmental Biology?' *Phronesis* 52.3 (2007), pp. 251–69.
Henry, Devin M., 'Understanding Aristotle's Reproductive Hylomorphism', *Apeiron* 39.3 (2006), pp. 257–87.
Hill, Rebecca, *The Interval: Relation and Becoming in Irigaray, Aristotle, and Bergson* (New York: Fordham University Press, 2012).

Holmes, Brooke, *Gender: Antiquity and Its Legacy* (Oxford: Oxford University Press, 2012).
Horowitz, Maryanne Cline, 'Aristotle and Woman', *Journal of the History of Biology* 9.2 (1976), pp. 183–213.
Irigaray, Luce, 'Body against Body: In Relation to the Mother', in *Sexes and Genealogies*, trans. Gillian C. Gill (New York: Columbia University Press, 1993), pp. 7–22.
Irigaray, Luce, 'How to Conceive (of) a Girl', in *Speculum of the Other Woman*, trans. Gillian C. Gill (Ithaca, NY: Cornell University Press, 1985).
Irigaray, Luce, 'The Power of Discourse', in *This Sex Which is Not One*, trans. Catherine Porter (Ithaca, NY: Cornell University Press, 1985), pp. 68–85.
Jaeger, Werner, 'The Pneuma in the Lyceum', *Hermes* 48.H.1 (1913), pp. 29–74.
Jaulin, Annick, 'Making Sex, Thomas Laqueur and Aristotle', *Femmes, Genre, Histoire* 14 (2001), pp. 195–205.
Johnson, Monte Ransom, *Aristotle on Teleology* (Oxford: Clarendon Press, 2005).
Jones, Barrington, 'Aristotle's Introduction of Matter', *The Philosophical Review* 83.4 (1974), pp. 474–500.
Kahn, Charles, 'Aristotle on Thinking', in Martha Craven Nussbaum and Amélie Oksenberg Rorty (eds), *Essays on Aristotle's De Anima* (Oxford: Clarendon Press, 1980).
Katayama, Errol G., *Aristotle on Artifacts: A Metaphysical Puzzle* (New York: State University of New York Press, 1999).
Keul, Eva, *The Reign of the Phallus* (Berkeley: University of California Press, 1993).
King, Helen, *Hippocrates' Woman: Reading the Female Body in Ancient Greece* (New York: Routledge, 1998).
King, Helen, *The One-Sex Body on Trial: The Classical and Early Modern Evidence* (Farnham: Ashgate, 2013).
King, Hugh R., 'Aristotle without Prima Materia', *Journal of the History of Philosophy* 17 (1956), pp. 370–89.
Kirkland, Sean D., 'Dialectic and Proto-Phenomenology in Aristotle's *Topics* and *Physics*', *Proceedings of the Boston Area Colloquium of Ancient Philosophy* 29.1 (2014), pp. 185–213.
Kitto, H. D. F., *The Greeks* (Harmondsworth: Pelican Books, 1951).
Kosman, L. Aryeh, 'Animals and Other Beings in Aristotle', in A. Gotthelf and J. Lennox (eds), *Philosophical Issues in Aristotle's Biology* (Cambridge: Cambridge University Press, 1987), pp. 360–91.
Kosman, L. Aryeh, 'Male and Female in Aristotle's *Generation of Animals*', in James G. Lennox (ed.), *Being, Nature, and Life in Aristotle: Essays in Honor of Allan Gotthelf* (Cambridge: Cambridge University Press, 2010), pp. 147–67.

Kotzin, Rhoda, 'Aristotle's Views on Women', *American Philosophical Association Newsletter on Philosophy and Feminism* 88.3 (1989), pp. 21–5.
Lang, Helen, *The Order of Nature in Aristotle's Physics: Place and the Elements* (Cambridge: Cambridge University Press, 1998).
Lange, Lynda, 'Woman is Not a Rational Animal: On Aristotle's Biology of Reproduction', in Sandra Harding and Merrill B. Hintikka (eds), *Discovering Reality* (Boston: D. Reidel Publishing, 1983), pp. 1–15.
Laqueur, Thomas, *Making Sex: Body and Gender from the Greeks to Freud* (Cambridge, MA: Harvard University Press, 1990).
Lennox, James, 'Aristotle's Biology', in Edward N. Zalta (ed.), *The Stanford Encyclopedia of Philosophy* (Spring 2014 edition), <http://plato.stanford.edu/archives/spr2014/entries/aristotle-biology/> (last accessed 13 November 2018).
Lesher, James H., 'Aristotle on Form, Substance, and Universals: A Dilemma', *Phronesis* 16 (1971), pp. 169–78.
Lesky, Erna, *Die Zeugungs und Vererbungslehren der Antike und ihr Nachwirken* (Wiesbaden: Akademie der Wissenschaften und der Literatur in Mainz, 1950).
Leunissen, Mariska, '"Crafting Natures": Aristotle on Animal Design', *Philosophic Exchange* 41.1 (2011), Art. 3.
Leunissen, Mariska, *Explanation and Teleology in Aristotle's Science of Nature* (Cambridge: Cambridge University Press, 2010).
Lewis, Eric, 'Introduction', in *Alexander of Aphrodisias: On Aristotle's* Meteorology 4, trans. Eric Lewis (Ithaca, NY: Cornell University Press, 1996), pp. 1–60.
Lewis, Frank, 'Plato's Third Man Argument and the "Platonism" of Aristotelianism', in J. Bogen and J. E. McGuire (eds), *How Things Are: Studies in Predication and the History and Philosophy of Science* (Dordrecht: Springer, 1985), pp. 133–74.
Lloyd, A. C., 'Aristotle's Principle of Individuation', *Mind* 79.316 (1970), pp. 519–29.
Lloyd, G. E. R., *Science, Folklore, Ideology: Studies in the Life Sciences in Ancient Greece* (Cambridge: Cambridge University Press, 1983).
Lo Presti, Roberto, 'Informing Matter and Enmattered Forms: Aristotle and Galen on the "Power" of the Seed', *British Journal for the History of Philosophy* 22.5 (2014), pp. 929–50.
Long, Christopher P., *Aristotle on the Nature of Truth* (Cambridge: Cambridge University Press, 2010).
Loraux, Nicole, *The Children of Athena: Athenian Ideas about Citizenship and the Division between the Sexes*, trans. Caroline Levine (Princeton: Princeton University Press, 1993).

Loux, Michael J., 'Form, Species, and Predication in Metaphysics Z, H, and Θ', *Mind* 88.349 (1979), pp. 1-23.

Mahowald, Mary, *The Philosophy of Women* (Indianapolis: Hackett, 1983).

Manuli, Paola, '*Donne Mascoline, Femmine Sterili, Vergini Perpetue. La Ginecologia Greca Tra Ippocrate e Sorano*', in Silvia Campese, Paola Manuli and Giulia Sissa, *Madre Materia: Sociologia e Biologia Della Donna Greca* (Turin: Boringhieri, 1983), pp. 13-79.

Manuli, Paola, '*Fisiologia e Patologia del Femminile Negli Scritti Ippocratici Dell' Antica Ginecologia Greca*', in Mirko D. Grmek (ed.), *Hippocratica: Actes du Colloque Hippocratique de Paris 1978* (Paris: Editions de CNRS, 1980), pp. 393-408.

Mayhew, Robert, *The Female in Aristotle's Biology: Reason or Rationalization* (Chicago: University of Chicago Press, 2004).

Mendelsohn, Daniel, 'An Odyssey: Fathers, Sons, and a Homeric Epic', *New Yorker* (24 April 2017), pp. 54-9. Quoting Homer, *Odyssey* I.215.

Morsink, Johannes, 'Was Aristotle's Biology Sexist?' *Journal of the History of Biology* 12.1 (1979), pp. 87-112.

Northwood, Heidi, 'Disobedient Matter: The Female Contribution in Aristotle's Embryology', *Epoché* 9.2 (2005), pp. 345-58.

Nussbaum, Martha Craven, 'Aristotle, Feminism, and Needs for Functioning', in Cynthia Freeland (ed.), *Feminist Interpretations of Aristotle* (University Park: Pennsylvania State University Press, 1998), pp. 248-59.

Nussbaum, Martha Craven, 'Aristotle on Teleological Explanation', in *Aristotle's De Motu Animalium* (Princeton: Princeton University Press, 1978), pp. 59-106.

Nussbaum, Martha Craven, 'The Role of *Phantasia* in Aristotle's Explanations of Action', in *Aristotle's De Motu Animalium* (Princeton: Princeton University Press, 1978), pp. 221-69.

Ober, Josiah, *Mass and Elite in Democratic Athens: Rhetoric, Ideology and the Power of the People* (Princeton: Princeton University Press, 1991).

Oele, Marjolein, 'Heidegger's Reading of Aristotle's Concept of *Pathos*', *Epoché* 16.2 (2012), pp. 389-406.

Oele, Marjolein, 'Passive Dispositions: On the Relationship between πάθος and ἕξις in Aristotle', *Ancient Philosophy* 32.2 (2012), pp. 351-68.

Okin, Susan Moller, *Women in Western Political Thought* (Princeton: Princeton University Press, 1978).

Owen, G. E. L., 'Dialectic and Eristic in the Treatment of the Forms', in G. E. L. Owen (ed.), *Aristotle on Dialectic: The Topics. Proceedings of the Third Symposium Aristotelicum* (Oxford: Clarendon Press, 1968), pp. 103-25.

Panofsky, Dora, and Erwin Panofsky, *Pandora's Box: The Changing Aspects of a Mythical Symbol* (New York: Pantheon Books, 1956).

Park, Katherine, and Robert A. Nye, 'Destiny is Anatomy', *The New Republic* (18 February 1991), pp. 53–7.

Pellegrin, Pierre, 'Aristotle: Zoology without Species', in Allan Gotthelf (ed.), *Aristotle on Nature and Living Things: Philosophical and Historical Studies*, trans. Anthony Preus (Pittsburgh: Mathesis Publications, 1985), pp. 95–116.

Pinto, Rhodes, '"All Things Are Full of Gods": Souls and Gods in Thales', *Ancient Philosophy* 36.2 (2016), pp. 243–61.

Pohlenz, M., *Hippokrates und die Begrundung der wissenschaftlichen Medizin* (Berlin: de Gruyter, 1938).

Posner, Richard, *Sex and Reason* (Cambridge, MA: Harvard University Press, 1992).

Preus, Anthony, 'Science and Philosophy in Aristotle's *Generation of Animals*', *Journal of the History of Biology* 3.1 (1970), pp. 1–52.

Rabinowitz, Nancy Sorkin, *Anxiety Veiled: Euripides and the Traffic in Women* (Ithaca, NY: Cornell University Press, 1993).

Rashed, Marwar, 'A Latent Difficulty in Aristotle's Theory of Semen: The Homogeneous Nature of Semen and the Role of the Frothy Bubble', in Andrea Falcon and David Lefebvre (eds), *Aristotle's Generation of Animals: A Critical Guide* (Cambridge: Cambridge University Press, 2018), pp. 108–29.

Reeve, C. D. C., *Substantial Knowledge: Aristotle's Metaphysics* (Bloomington, IN: Hackett, 2000).

Robinson, Howard M., 'Prime Matter in Aristotle', *Phronesis* 19.1 (1974), pp. 168–88.

Salamon, Gayle, 'Sameness, Alterity, Flesh: Luce Irigaray and the Place of Sexual Undecideability', in Athena Athanasiou and Elena Tzelepis, *Rewriting Difference: Luce Irigaray and 'the Greeks'* (Albany: State University of New York Press, 2010), pp. 191–201.

Salmieri, Gregory, 'Something(s) in the Way(s) He Moves: Reconsidering the Embryological Argument for Particular Forms in Aristotle', in Andrea Falcon and David Lefebvre (eds), *Aristotle's Generation of Animals: A Critical Guide* (Cambridge: Cambridge University Press, 2018), pp. 188–206.

Salque, Mélanie, Peter I. Bogucki, Joanna Pyzel, Iwona Sobkowiak-Tabaka, Ryszard Grygiel, Marzena Szmyt and Richard P. Evershed, 'Earliest Evidence for Cheese Making in the Sixth Millennium in Northern Europe', *Nature* 493 (24 January 2013), pp. 522–5.

Scaltsas, Theodore, 'Aristotle's "Second Man" Argument', *Phronesis* 38.2 (1993), pp. 118–36.

Schleiner, Winfried, 'Early Modern Controversies about the One-Sex Model', *Renaissance Quarterly* 53.1 (2000), pp. 180–91.

Schofield, Malcolm, 'Ideology and Philosophy in Aristotle's Theory of Slavery', in Richard Kraut and Steven Skultety (eds), *Aristotle's Politics: Critical Essays* (Lanham, MD: Rowman & Littlefield, 2005), pp. 91–120.

Sharkey, Sarah Borden, *An Aristotelian Feminism* (Cham, Switzerland: Springer, 2016).

Shaw, Michael M., 'Aither and the Four Roots in Empedocles', *Research in Phenomenology* 44.2 (2014), pp. 170–93.

Sissa, Guilia, *Greek Virginity* (Cambridge, MA: Harvard University Press, 1990).

Solmsen, Friedrich, 'Aristotle and Prime Matter: A Reply to Hugh R. King', *Journal of the History of Ideas* 19.2 (1958), pp. 243–52.

Solmsen, Friedrich, *Aristotle's System of the Physical World: A Comparison with His Predecessors* (Ithaca, NY: Cornell University Press, 1960).

Solmsen, Friedrich, 'Nature as Craftsman in Greek Thought', *Journal of the History of Ideas* 24.4 (1963), pp. 473–96.

Solmsen, Friedrich, 'The Vital Heat, the Inborn Pneuma and the Aether', *The Journal of Hellenic Studies* 77.1 (1957), pp. 119–23.

Sorabji, Richard, 'Aristotle on the Role of Intellect in Virtue', in Amélie Oksenberg Rorty (ed.), *Essays on Aristotle's Ethics* (Berkeley: University of California Press, 1980), pp. 201–20.

Stewart, Andrew, 'Rape?' in Ellen Reeder (ed.), *Pandora: Women in Classical Greece* (Baltimore: The Walters Art Gallery, and Princeton University Press, 1995), pp. 74–90.

Studtmann, Paul, 'Living Capacities and Vital Heat in Aristotle', *Ancient Philosophy* 24.2 (2004), pp. 365–79.

Tkacz, Michael W., 'Albertus Magnus and Recovery of Aristotelian Form', *The Review of Metaphysics* 64.4 (2011), pp. 735–62.

Tress, Daryl McGowan, 'Aristotle against the Hippocratics on Sexual Generation: A Reply to Coles', *Phronesis* 44.3 (1999), pp. 228–41.

Tress, Daryl McGowan, 'The Metaphysical Science of Aristotle's *Generation of Animals* and Its Feminist Critics', *Review of Metaphysics* 46.2 (1992), pp. 307–41.

Trott, Adriel M., *Aristotle on the Nature of Community* (Cambridge: Cambridge University Press, 2014).

Tuana, Nancy, 'Aristotle and the Politics of Reproduction', in Bat-Ami Bar On (ed.), *Engendering Origins: Critical Feminist Readings in Plato and Aristotle* (Albany: State University of New York Press, 1994), pp. 189–207.

Whitbeck, Caroline, 'Theories of Sex Difference', in Carol C. Gould and Marx W. Wartofsky (eds), *Women in Philosophy* (New York: G. P. Putnam's Sons, 1976), pp. 54–80.

Whiting, Jennifer E., 'Form and Individuation in Aristotle', *History of Philosophy Quarterly* 3.4 (1986), pp. 359–77.

Whiting, Jennifer E., 'Metasubstance: Critical Notice of Frede-Patzig and Furth', *The Philosophical Review* 100.4 (1991), pp. 607–39.

Wiggins, David, 'Deliberation and Practical Reason', in Amélie Oksenberg Rorty (ed.), *Essays on Aristotle's Ethics* (Berkeley: University of California Press, 1980), pp. 221–40.

Wilshire, Donna, 'The Uses of Myth, Image, and the Female Body in Re-visioning Knowledge', in Alison M. Jaggar and Susan Bordo (eds), *Gender/Body/Knowledge: Feminist Reconstructions of Being and Knowing* (New Brunswick, NJ: Rutgers University Press, 1989), pp. 92–114.

Witt, Charlotte, 'Aristotle's Essentialism Revisited', *Journal of the History of Philosophy* 27.2 (1989), pp. 285–98.

Witt, Charlotte, 'Form, Normativity, and Gender in Aristotle: A Feminist Perspective', in Cynthia Freeland (ed.), *Feminist Interpretations of Aristotle* (University Park: Pennsylvania State University Press, 1998), pp. 118–37.

Witt, Charlotte, 'Form, Reproduction, and Inherited Characteristics in Aristotle's *Generation of Animals*', *Phronesis* 30.1 (1985), pp. 46–57.

Witt, Charlotte, *Substance and Essence in Aristotle: An Interpretation of* Metaphysics VII–IX (Ithaca, NY: Cornell University Press, 1989).

Woods, Michael J., 'Problems in *Metaphysics* Z, Chapter 13', in J. Moravcsik (ed.), *Aristotle: A Collection of Critical Essays* (South Bend, IN: University of Notre Dame Press, 1968), pp. 215–38.

Yates, Velvet L., 'Biology is Destiny: The Deficiencies of Women in Aristotle's Biology and Politics', *Arethusa* 48.1 (2015), pp. 1–16.

Zeitlin, Froma, *Playing the Other: Gender and Society in Classical Greek Literature* (Chicago: University of Chicago Press, 1996).

Index Locorum

AESCHYLUS
Eumenides
1–3	131
151–2	127
211–13	127
261	127
284	127
321–96	127
421	127
434–5	127
608	127
625–8	128
632–5	128
640–4	129
658–66	129
736	129
825–32	130
885–900	130
944–8	129
956–7	130

Seven Against Thebes
956–9	128

ALEXANDER OF APHRODISIAS
On Aristotle's Meteorology
186	19–21, 158
208	21, 158

ANAXAGORAS
Fr. 1	135
Fr. 12	135

APOLLODORUS
Library 3.4.3	122

ARISTOTLE
Categories
12a1–17	209
13a19–21	209

De Anima
411a8–9	134
412a20–1	110
412b6–9	111
413a8–9	111
413b1–3	111
414a4–7	72
414a13–14	72
414a26–7	112
415a24–7	167
415a27–b1	239
415b9–12	111, 176
416a10–19	175
416a19–24	113
416a26–9	156
416a34–b8	115
416b28–9	185
417a23–8	115–16
417b2–9	115, 117
417b31–4	116
424a18–29	58
426a6–11	216
429a23–8	113
432a1–2	113

De Caelo
270a13–b30	53
289a20–4	180
295a29	136
301a7–11	94

De Caelo (cont.)
302a15–19	147
303a25–7	147
304b12–14	94
304b25–7	147
305a5	95
305a31–3	95, 147
306b16–21	94
307a31–b4	104
308b10–28	94
310a31–b15	96
311a19–b29	96
312a16–23	96–7
312a30–2	92
312b20–3	94

De Motu Animalium
701a35–6	177
701b17–25	177
703a12–15	175
703a29–b2	170
703b19–20	177
703b25–6	170

Generation of Animals
716a13–31	30, 41, 227
722a3–7	169
722b1–3	169
724a7–10	31
724a17–20	30
724a35–b6	30
724b7–10	206
724b22–726a6	48, 145, 174, 225
726b2–11	139, 143
726b20–1	35
727a3–6	143
727a31–b5	53, 143, 145, 231
728a16–21	144
728a26–30	143
728b21–5	31
729a11–20	184, 228
729a32–3	41
729b5–21	174, 213–14
730a27–9	177
730a35–730b31	216–17
730b10–15	60
730b15–19	35
730b35	66
731a25	218
731b25–8	35
732a16–19	165, 182
733a11	150
733b25–734a16	166–7
734a34–b3	168–9
734b12–17	170, 219
734b18–19	173
734b22–6	170, 220
734b28–735a4	166, 172
735a1–2	220
735a12–14	173
735a22–5	175, 193
735b4–6	178
735b10–15	178
735b34–5	178
736a1–2	178
736b34–737a6	166, 178–9, 183
737a6–16	54, 60, 174, 184, 188, 229
737a27–8	188
738a31–6	145
738b20–6	3, 220, 229
739b20–9	182, 230
739b35	221
740a24–31	221
740b4–5	221
740b19–20	221
740b24–741a2	4, 222
741a17–18	185
741b15–21	193
742a19–b8	232
742b7–8	232
743a4–743b24	153, 159, 224
744a36–b1	233
744b15–26	233
747a18–19	180
748a8–15	11
749b3–9	31
750b4–5	31
762a18–31	58, 134, 177, 180, 224
762b9–18	58
763b23–4	191
763b28–764a22	191
764a1–765a4	169, 174
765a35–b6	191
765b9–20	4, 145, 192
766a14–15	192
766a17–20	166, 193, 206
766a21–9	188, 193, 206–7, 209
766a30–b3	193, 207
766b12–18	31, 194
766b19–25	188

INDEX LOCORUM 257

766b27–767a12	195	1042a26–7	91
767a20–7	195	1046a19–28	216
767b7	46	1047a4–7	39
767b20–4	196	1049b5–10	109
768a1–6	46, 196, 208	1050b2	66
768a11–14	198, 200	1055a33	207
768a29–31	201	1055b1–2	209
768b12–29	201–2	1057a18–19	209
769b11–30	188	1058a29–36	209
770b13–17	41	1058b21–5	31, 210
771a31–3	231	1071a18–24	70, 109
771b18–28	231	1072b3–4	35
772a16–29	231	1078a5–7	31
775a4–15	189–90		
784b4–7	156, 182	*Meteorology*	
789b8–15	226	354b33–355a7	133
		355a5	156
History of Animals		355b10	182
491b26–34	189	378b21–5	150
498a32–b3	189	379a1	146
532b30–533a15	189	379a11–12	156
		379a16–17	156
Magna Moralia		379a24–6	156, 182
1210a15–21	148	379b11	150
		379b18–20	147, 158, 182
Metaphysics		379b24–6	148, 182
983b20–4	133	379b33–5	148
984a5	134	380a2	182
985a31–3	135	380a9	150
990b22–991a2	70	380a20–6	151
1014b26–35	82	380a32	151
1028a20–30	70	380a34	150
1028b26–31	70	380b12–23	151
1029a9–b12	67, 69, 91–2	380b31–5	151
1030a17–b5	66	381a24	151
1031b7–20	68, 69	381a29	150
1032b1–2	66	381b7–9	185
1032b14–15	70	381b24–5	150
1033a30–1	70	382a28–30	153
1033b17	66	383a13–17	101
1034a2–8	65–6	383a19–20	158
1034a21–30	65	383b19–20	101
1034a35–b8	65	383b24–5	178
1034b10–15	65	384a4	151
1035b14–21	70	384a12–17	151
1036a28	66	384a21–4	151
1037a10	70	384a35–384b1	151
1037a27–b7	66, 70	385a2–6	151
1038b8–12	66	385a19–20	151
1040a5–7	66	388a21–4	147
1041b4–9	66, 72	388a31–3	158

Meteorology (cont.)
388b9–10	158
389a19–20	178
390b3–19	159, 165

Nicomachean Ethics
1109b18–20	46
1112b12	113
1142b31–3	39
1178a24–b8	177

On Breath
483a13	181
483a30–5	180
485a28–485b8	154

On Generation and Corruption
314a8–12	97
314b28–9	97
315b22–4	97
316b29–35	98
317a3	98
317a19–26	99
318a32–b9	100
318b14–18	101
319b8–11	102
319b14–24	102
320a1–4	103, 107
320b22–3	97
322b16–17	101
323b3–9	104
323b18–24	104
323b29–324a14	105
324a11	152
324a16–19	105
325a26–9	98
325b18–22	98
326a1–10	99
326a24–9	99
329a25–7	103
329b23–30	104
331a26–331b2	104
335b30–1	40
344a1	136

On Respiration
473a2–13	181, 186
474a28–b3	175
474b10–12	153
476a15–18	153
479a8–9	153
480b4–5	180

On Sleep and Waking
456b5–6	185
458a27	182

On Youth and Old Age (Juv.)
464b14–16	158
465a14–16	156
466a20–4	150
468b28–469a9	158, 175
469b7–8	182
469b10–19	181
469b15–20	175
470a5–6	153
470a20–5	153, 182
473a11	182
474b13	182
478a16	182
483a13	181
483a30–5	180

Parts of Animals
646a18–21	147
648a8–14	151, 153
648a27	182
648a37	154
648b4–11	148
649a1–2	182
649a16–17	154
649a20–9	152–3
649b12–13	155
649b24–35	155–6
650a2–9	158, 184
650a15	182
650a23–7	184
651a11–12	182, 185
652a35–7	136, 153
652b8–16	166, 181
663a29–34	190
663b32–664a3	190
668b11–15	190
670a22–5	175
672b27–32	190
676a6	184
684a32–b1	189
697a15–b14	189

Physics
187a12–26	91, 134
189b33–190a4	83
190a12–21	83
190a31–6	84
190b4	84

190b20	84	*Gland*	
190b29–191a7	84, 205	17	140
191a9–12	85		
192a4–6	87, 205	*Nature of the Child*	
192a9–15	161	1	139
192a20–4	20	3	139
192a25–33	89, 205	4	139
192b14–21	109	6	139–40
192b33	70		
193b6–7	109	*Nature of Man*	
193b11–12	199	I.1–18	132
193b18–19	199	II.2–8	132
198a25–9	112–13	II.12–14	132
200b12	3	II.17–29	132
202b3–8	216	III.19–23	132
202b17–21	216	IV.1–20	132
213a1–3	94	IX.4–10	133

Politics

		Places in Man	
1254b16–24	234	42	140
1259b28–9	234		
1260b5–7	234	*Regimen* I–III *(Vict.)*	
1309b23–34	46	I.3	156
		I.9	182

Problems

		I.10.1–26	133, 180
860a34	182	I.25.8	182
883a7	182	I.27.1–9	138
909b15–16	182	I.27.18–25	138
949b5	182	I.29	182
		II.62	182

HESIOD

PROCLUS

Theogony

Commentary on Plato's Parmenides

91–6	125	792.5–19	74
116–32	122	794.3–20	74
590–612	124	795.26–7	74
		797.19–32	74

HIPPOCRATICS

		844.1–5	74
Airs, Water, Places		883.1–6	74
10	138	883.20–2	74
		891.32	74
Aphorisms		897.37–898.1	74
I.14	182	949.13–18	74
I.15	182	950.5–7	74
V.63	182		

SOPHOCLES

Oedipus Rex

Diseases (Morb.)

I.11	182	569	128

General Index

Achilles, 126
actuality, actual
   form as, 36–9, 58, 62, 72, 77, 86, 94, 105, 152, 155, 160, 162, 172, 186, 216
   in generation, 99, 101, 149, 160, 162, 171–3, 179, 183, 190, 195, 197, 216, 223, 225
   knowledge, 66, 72, 76, 78
   male as, 34, 58n24
   of material, 78
   movements in resemblance, 45, 198, 201–3
   semen's work for, 110, 166, 168, 194, 200–1, 215–16
   soul as, 71–2, 110–20, 168, 176–7
   of substance, 33–4, 46, 89, 200, 220–1
Adelman, Janet, 19n45
Adorno, Theodor W., 23
Aeschylus, 125, 128 31, 136, 144
aether, 36, 53n5, 53–7, 56n17, 177, 179, 184
Agamemnon, 125–6, 128
agent, 5, 98–101, 104–5, 110, 114, 117, 153, 167
Agnodice, 19
air, 54n10, 55–6, 94, 96, 103, 106–9, 132–5, 139, 147–54, 163, 177–81, 184, 224–5
Al-Farabi, 54
Albertus Magnus, 55, 76
Albritton, Rogers, 64n48
Alexander of Aphrodisias, 149–50, 150n19, 158
alteration, 97–9, 102–9, 115
Amazon, 128
Anaxagoras, 132, 134–5, 142, 149, 191
Anaximander, 91, 134–5, 141
Anaximenes, 134–5
Annas, Julia, 71n66
Antigone, 126
Antiphon, 2, 92
Aphrodite, 177
Apollo, 126–9, 131, 144

architecture, 171, 215–16, 219
Artemis, 126, 131
artifice, artefacts, τέχνη
   compared with nature, 2–4, 12, 28, 110–11, 118–19, 172–3, 174n84, 176, 212, 225, 237–8
   matter in, 47–8, 80, 82, 83n9, 85, 88n26, 89–90, 92
   tools in, 217, 222
Asklepios, 139, 140n92
Athamantas, 122
Athena, 121–2, 127, 129–30
Athens, 13–14
atom, atomism, atomists, 95, 98
automaton *see* puppetry, puppets
Avicenna, 54

Balme, D. M., 55–6, 62, 77, 198n30, 201n38
Baracchi, Claudia, 2n3
Barnes, Jonathan, 8
Bartoš, Hynek, 140, 157n43, 180, 182n121
base/superstructure model, 6, 25–6
bats, 189
Beare, J. I., 182
Beauvoir, Simone de, 5
belly *see* stomach
Bianchi, Emanuela, 6, 22, 39–42, 44, 49, 79, 212n1
binary, 3–5, 20, 25
birds, 189
Bleier, Ruth, 28n6
Blok, Josine, 13–15
blood
   amount of, 188, 190, 192, 238
   formation of, 63, 89, 102, 108, 144–6, 150, 167, 183–6, 193, 207
   in Greek mythology, 121, 124, 126–7, 129–30

Hippocratic view of, 131–2, 139, 141, 143
as homoeomerous part, 159, 165, 178
menses formed from, 33, 43, 87, 159, 173, 221
role in courage of, 153
semen formed from, 32, 43–4, 57–8, 152, 154–5, 162, 170
body
  contrasted to mind, 25, 237
  difference between male and female, 11, 30–2, 41
  father's, 55
  health of, 56–7, 117
  living, 40, 48, 156–7
  models of sexed, 5–6, 17–23
  organisation of, 34, 43, 61, 72, 133, 176–7
  relation to soul of, 3–4, 36, 40, 64, 109–16, 118, 168, 186, 237
  sickness of, 132, 139
Bogen, James, 207
Bos, Abraham P., 38, 53, 194n22
Bostock, David, 68n60
Boys-Stones, George, 198n29
breath, πνεῦμα, 41, 56n17, 139, 141, 143, 152, 158, 177–81, 181n112, 220, 225–6
  as constitutive of semen, 50–1, 55–57, 60–61, 166
  as spiritual principle, 53, 54n10
builder, building, 43, 59, 118, 171, 215–16, 219
Butler, Judith, 6–8, 10, 15, 18, 79, 239
Byrne, Christopher, 93

Campese, Silvia, 125
capacity *see* potentiality
carpenter, 47–8, 74, 110, 163, 213–19
Carson, Anne, 136–7
change, 135
  accidental, 83–6, 99
  as alteration, 86, 95, 97, 99–100, 102–4, 108–9, 115, 160
  between contraries, 115–17
  elemental, 85, 93, 95, 100–8, 134, 136–7, 141, 147, 149–50, 152, 162
  internal, 39–40, 47
  material's role in, 41, 48, 153–4, 156
  in resemblance, 46
  substantial, 24, 45, 52, 60–1, 63, 69, 73, 81, 83–6, 97–100, 131–2, 160
  substratum, 83–7, 92
Charlton, William, 94, 95n54, 103, 108, 212
Clytemnestra, 125–7, 130–1
Code, Alan, 66n51, 69n62, 86n21
Cohen, Sheldon, 52, 84, 93

cold
  elemental force of, 63, 89, 93, 97, 101–2, 104–8, 150–4, 156
  female associated with, 19, 29, 33, 132–3, 144–8, 190–2, 194, 222–3
  Hippocratic view of, 32–3, 138–40
  in homoeomerous parts, 158–61, 172, 220, 232–3
  intelligence associated with, 153
  Pre-Socratic views of, 134
  in resemblance, 202
  in sexual differentiation, 45
  as tool of soul, 4, 52, 222–3
Cole, Eva Browning, 28
Colebrook, Claire, 39
Coles, Andrew, 29n8, 32, 53n8, 58, 77, 181n111
concoct, concoction
  blood formed through, 63, 145, 152, 175, 183–4
  definition of, 146–8, 154–8
  difference between male and female as degrees of, 4, 19, 31–3, 45, 65, 110, 145, 154–8, 177–8, 190–5, 197, 202–3, 206–8, 214, 221, 231, 233, 239
  menses formed by, 87, 223
  nourishment formed by, 57, 62
  semen formed by, 144, 146, 152, 155–6, 160, 162, 170, 172, 183–4
  vital heat as cause of, 148, 154
Connell, Sophia M., 9, 12, 13, 39, 41–9, 58–61, 176, 180, 194n21, 195n25, 198n29, 199n33, 199n34, 203n42, 212, 218
contrary, contraries, contrarieties, 3, 21, 115, 117
  active and passive, 118, 146, 148–53, 158
  compared with contradictories, 209–10
  in elemental change, 99, 103–8, 132, 149–52
  form and matter as, 208–10, 236
  form and privation as, 205
  hierarchical nature of, 30, 38, 41–2, 44n73, 49
  male and female as, 30–1, 204–10
  material, 31, 208, 210
  in substantial change, 73, 87
cooking, 195
Cooper, John M., 32n27, 62n43, 165, 198n29
craft analogy
  to nature, 1–3, 39–40, 42–3, 47–9, 78–80, 118, 166, 199, 241
  to sexual generation, 212–18, 220, 222–6, 228, 234
  see also artifice, artefacts
Creon, 128

D'Hoine, Pieter, 71n66
Dancy, Russell, 93
Dean-Jones, Lesley, 20n52, 29n9, 51n1, 54, 61–2, 121, 138, 141
deformity, 187, 188n3, 189n8, 189–90, 207
Demeter, 130
Democritus, 98, 105, 162–3, 191
departure from type, 208
Deslauriers, Marguerite, 31–32, 41n57, 61n35, 188n2, 198n31, 210
destruction, decay, 81, 95–6, 115, 122, 124–5, 133, 135, 141, 153–4, 156–7, 159, 175, 192–4
Diogenes of Apollonia, 136
Dionysus, 121–2, 129, 131
doctor, doctoring, 3, 40, 117, 226
dolphins, 189
dry, 101, 104–8, 132–3, 136, 138, 140, 147–55, 184, 220, 232
DuBois, Page, 28

earth, 42, 89, 96, 100–1, 106, 122, 130–5, 141, 147–54, 163, 178, 183–4, 224–5, 227–8
efficient cause *see* moving cause
ek (ἐκ) relation, 84–5, 158
elemental forces, powers
 as constituent parts of elements, 89, 159–61, 180
 in elemental generation, 81, 100–8, 149–50
 Hippocratic view of, 133–5, 140–1
 Pre-Socratic views of, 134–6
 in sense perception, 58
 in sexual reproduction, 110, 144, 146–8, 152, 156, 183, 223, 239
elements, elemental
 common matter of, 81, 92–4
 elemental forces in, 133, 148
 generation of, 80, 91–109, 149–50, 160–1, 177–8
 Hippocratic view of, 133–5
 as material, 63, 89, 110, 112, 144, 159, 206, 239
 Pre-Socratic views of, 133–41
Elshtain, Jean Bethke, 28n7
embryo
 as seed, 86
 growing in female, 214–16, 221
 material in, 54, 222–3, 230, 233
 menses in, 186, 232
 semen's role in causing, 62, 170–1, 173–4, 184, 190–1, 194, 200, 202–3, 213, 217, 219, 229, 231–2
 sex determination and trait inheritance in, 193–7, 201, 206, 208
 soul in, 55, 114, 160, 167–8, 188, 236

Empedocles, 34, 54, 98, 132, 135–6, 149, 162–3, 179, 191
essence
 of female, 18, 21, 41
 as form, 22, 64, 66–75, 137, 148, 205
 soul as, 72, 111, 176
eunuch, 187, 193, 209
Euripides, 126

fat, 61–2, 145, 231
father, 76, 114, 160, 164, 166, 170–2, 174, 176, 197, 201, 219, 226, 236
female
 associated with body, 30
 capacity to be generated in, 30, 33–4, 39–40, 47, 214–16, 227–8
 coldness of, 29, 145–6, 148, 152, 239
 container view of, 28, 30, 34
 contribution, 32–4, 36, 43–4, 54, 110, 143–4, 146, 165, 182–3, 187, 214
 as deformed or mutilated, 187–90
 gods, 120–31
 incapacity of, 4, 10, 21, 33, 47, 143–6, 191, 193–6, 207
 incomplete nature of, 38, 46
 independence from male, 59–60
 as irrational, emotional, 11, 30
 as material, 3–5, 9–10, 13, 17–18, 20, 22–3, 27, 29–30, 33, 35, 37–9, 41–4, 48, 54–5, 110, 158, 165, 214, 220–2, 225–6, 229, 238
 menses, 58, 63
 moisture in, 9, 137–41, 146, 148, 152, 195
 offspring, 32, 44
 as opposite to male, 15, 87, 204–9
 as passive, 45, 174
 principle distinguished from animal, 31, 46, 61
 role in resemblance, 19, 46, 187, 195–203
 role in sexual differentiation, 188–97
 seed, 29n9, 30–2, 44, 86, 139, 194
 as unpredictable, 40, 42, 236
female material, 34, 37, 41, 43–4, 72, 196, 207, 210, 229, 238
fetation, κυήμα, 36, 45, 118–19, 197
Fielding, Helen, 15n34
fig-juice, 184, 228–31
final cause, 1, 40, 47–8, 52–3, 55, 59, 62–3, 111–12, 114, 137, 161–3
fire
 contrasted with vital heat, 153, 181, 181n112, 165–6, 179–81
 as element, 132
 generation of, 100–1, 105–6, 108
 Hippocratic view of, 132, 138, 156–7

GENERAL INDEX 263

power of, 104, 122, 133, 152, 161, 180-1, 185
Pre-Socratic views of, 133, 135, 137
relation to form of, 96
as tool of soul, 112
as unlimited heat, 63, 150, 175, 182
fish, 189
Flemming, Rebecca, 21n56
flesh, 52, 63, 67, 86n21, 89, 139, 147, 150, 159, 169-72, 186
fluidity, 18-19, 21-2
food, 32, 57, 110, 113, 115-18, 150, 184-5, 233
form, formal principle, εἶδος, 92, 147, 153
   actual, actualising of, 160, 226
   as arrangement, 2, 97, 99, 112, 160, 163, 234, 238
   artificial, 2-4
   association of reason with, 36, 42
   as capacity, 207
   craft analogy view of, 39-40, 43, 47, 85-6, 171-2
   deformity of, 189
   difference, 22, 26, 31-3, 37, 240
   elemental, 96, 103-9, 150-1, 161
   in elemental change, 96, 140, 147-8
   essence as, 64-9, 71-2
   father's, 160, 166, 170, 179
   functional, 38, 63
   in generation, 28, 30, 36, 47-8, 51, 61, 63, 80-1, 83-4, 93, 110
   incapacity of, 190
   internalised heat as, 8, 154-5, 158-9, 179, 189
   male as, 4, 17, 27, 30, 34, 43, 62
   material basis of, 12-13, 24, 49, 58, 181, 187
   as moving cause, 4
   organising capacity of, 59, 64, 218
   Plato's view of, 64, 68-73, 75-6, 82
   potential, 160
   privation of, 87, 89
   relation to matter, 3-11, 15, 18, 20-1, 23-5, 40-2, 47, 51, 53-4, 79-80, 118-19, 163, 204, 237, 240
   semen's role as, 3, 35, 50, 144, 160, 162, 165, 170-1, 214, 221, 228
   separable principle of, 59
   in sexual differentiation, 191-7
   soul as, 4, 111-14, 163, 223-4
   species versus individual, 52, 62n43, 64n48, 64-78, 179, 198-9, 201, 203, 198, 201n38, 212, 223
   superiority over matter of, 5, 7-8, 10-11, 18, 21, 33, 35, 37-8, 43-5, 236
   vital heat's role as, 53, 58, 63, 110, 154, 161, 165, 179, 181n112, 220

Frede, Michael, 73n73
Frede, Michael and Günther Patzig, 67
Freeland, Cynthia A., 38-9, 181n6
Freudenthal, Gad, 56-58, 181n112, 185n138
friendship, 148n12
Furies, 124, 127-8, 130
Furth, Montgomery, 198n32, 51

Gaia, 121-5, 127, 130
Galen, 16, 18n45, 44
Gelber, Jessica, 43, 188n1, 199n33, 199-200, 201n38
generation, γένεσις
   analogies for, 58-60, 83, 85
   contrasted to alteration, 86, 95-100, 102-4, 108-9, 115, 160
   craft model of, 79, 82, 119, 172, 212-25, 227
   elemental, 81, 92-6, 98, 95-100, 102-3, 107-8, 136, 181
   female role in, 10, 13, 27, 30-1, 34, 47, 182, 207
   form in, 15, 24-6, 28, 47, 52, 62, 65, 73, 165, 190, 204, 240
   heat in, 146, 158, 183
   Hippocratic view of, 139-41, 143-4, 197
   male role in, 27, 33, 207
   material in, 12, 24-6, 28, 47-8, 52, 72, 77-80, 89, 125, 145, 190, 195, 204, 206, 224, 240
   menses, 50
   natural, 39, 62, 80, 82, 109-11, 114, 116-19, 158, 160, 172, 183, 204-5, 225, 237-9
   of parts, 63
   Platonic and Neoplatonic view of, 73-5, 91
   Pre-Socratic views of, 134-6
   principles of, 11, 30
   semen's role in, 12, 13, 50-1, 55-6, 118, 169, 174, 190, 229
   sexual, 197, 213-16, 230-1
   spontaneous, 57
Gill, Mary Louise, 23-4, 62-4, 64n48, 85, 86n21, 88, 90, 108n90, 160-1, 161n50, 198n29, 205n44, 207n50
Gotthelf, Allan, 189n6
Green, Judith M., 38
Grosz, Elizabeth, 6, 8, 25, 240
growth, 4, 28, 30, 34, 57-8, 133, 139, 221-3, 175, 180, 184-6, 221-3, 233-5
Guthrie, W. K. C., 93

Hades, 130, 135
haematogenous thesis, 43-4
Hansen, Ann Ellis, 139n88
health *see* doctor

heart, 32, 55, 57-8, 154, 167, 169-70, 175-6, 181, 193, 207, 221, 235
heat, hot 63, 132-4, 138-40, 159-61, 200
  in blood, 57, 146, 152-3
  compared with vital heat, 50, 53
  in concoction, 8, 16, 19-20, 143-5, 148, 151
  in elemental change, 101, 104-6, 147-9
  formal work of versus material work of, 173, 178, 187
  in generation, 50-2, 55, 143, 189, 193, 208, 210
  internal, 152, 154-9, 159-62, 170, 178-83, 186, 193, 237
  in semen, 166, 170, 172, 218, 226, 232-3
  in sexual differentiation, 190-1, 195
  of soul, 218, 220, 222-3
  of the sun, 228
Hegel, G. W. F., 14
Heidegger, Martin, 1, 2n3, 15n36, 23n60
Henry, Devin M., 32, 33n28, 44n73
Hephaestus, 122-3, 131
Hera, 122-3, 135
Heraclitus, 133, 136-7, 180
heredity *see* inheritance
Hermes, 122
Hesiod, 122-5, 128, 135, 137, 141, 143
Hippocratics
  on elements in the body, 132-4, 136-41
  on female contribution, 9, 30-1, 44, 120, 141, 143-4, 146
  on material basis of soul, 61, 72, 177, 180, 182
  one-sex model view of, 20n52
  pansomatist view of, 53
  preformationist view of, 28
  on women, 124, 138
Holmes, Brooke, 8, 19, 20n52, 22
homoeomerous bodies, 43, 63, 147, 150, 154, 159-61, 166, 168, 185, 220, 232
homogeneous parts *see* homoeomerous bodies
homunculus, 28n7, 176, 197
house, home-building *see* builder, building
householder, 145n4, 164, 227, 233-5
hylomorphism, 5, 37, 68n60, 77, 79, 81, 107, 135, 238, 241
hylozoism, 59-60

Ibn Bâjja, 54
ideology, 9, 33
inheritance, 71
  departure from type in, ἐξίσταται, 46-7
  from the mother, 45, 187-91, 197-203
  form's work in, 65-6, 187-8, 197-203
  straying from type in, παρεκβέβηκε, 46

inherited traits, 71, 189, 197-203, 210
Ino, 122
insects, 217
Iphigeneia, 125-7, 131
Irigaray, Luce, 6-7, 10-11, 15, 18, 23, 39, 42, 79, 238-9

Jaulin, Annick, 18n45
Jones, Barrington, 83n9, 88n26, 89

Kahn, Charles, 177
Keul, Eva, 28
King, Helen, 18n45, 19, 20n52, 21n56, 120, 124-6, 138, 140
King, Hugh R., 81-2, 87, 90, 93-4
Kirkland, Sean D., 189n8
Kosman, L. Aryeh, 33, 81, 86, 88n26, 174n80
Kronos, 123

Lang, Helen, 96n76, 161n50
Laqueur, Thomas, 16-18
Lee, H. D. P., 150n19, 153n31
Lefkowitz, Mary R., 120n3, 123n7, 124n12, 131n41
Lennox, James, 51n1
Lesher, James H., 64n48
Leucippus, 98
Leunissen, Mariska, 162n55, 164, 164n60
Lewis, Eric, 84, 108n88, 148-9
Lewis, Frank, 69
life, living
  body or organism, 54, 59, 61-2, 64, 111-14, 116-19, 134, 152, 155-8, 161, 183, 195, 207, 214, 221, 226, 232, 234
  divine source of, 54-5
  female contribution to, 30, 110, 146
  form in causing, 38, 63, 77-8
  function of, 234
  heat's role in causing, 50, 166, 179-84, 224
  male as source of, 5, 16, 19, 21, 34, 36, 50, 146, 190-1, 194, 197, 203, 206-8, 214
  material source of, 43, 47, 52, 56, 59, 77, 132, 148, 152, 157, 160, 170-81, 225-6
  mythological view of, 121-4, 128, 130-1, 135, 141
  role of heat in generating, 51, 56, 58, 60
  role of semen in generating, 53, 56, 110, 116, 145, 164-8, 171, 174, 185, 193, 200, 217, 219-20, 222, 224, 238
  soul as source of, 61, 110-14, 118, 143
Lloyd, A. C., 67n56
Lloyd, G. E. R., 9
Lo Presti, Roberto, 72

lobster, 189
Long, Christopher P., 83
loosening *see* relapse
Loraux, Nicole, 124
Loux, Michael J., 64n48, 69n63
luxury parts, 164

Mahowald, Mary, 28
male
  as ability to concoct, 145, 191, 193
  capacity to generate in another of, 16–17, 39–40, 47, 145–6, 227–8
  as complete form of species, 29
  as contrary to female, 204, 206–8
  contribution, 32, 43–4, 110, 165, 213, 229
  as form, 3–4, 145, 187, 190, 238
  as heat, 146, 158, 166, 170, 172, 183, 218, 226, 232–3
  as principle of movement, moving cause, 4, 32, 35, 143, 145, 177, 187, 213–19, 222
  as rational, 11, 14, 29–30
  relation to material of, 159, 187
  role in resemblance, 197–203
  role in sexual differentiation of, 191–7
  seed, 29
  as soul, 3–4, 17, 34–6, 47, 55
  superiority of, 15, 18, 21, 33, 35–36, 45, 55–6, 141, 153
Manuli, Paola, 120
master, mastery, 41, 44–5, 146–8, 153–8, 170, 196, 198, 208
material, matter, ὕλη
  as acted upon, 150–1, 204
  building block notion of, 98–9, 101, 104
  causal power of, 237
  constitutive, 80–81
  craft versus nature view of, 212–24
  definition of, 7, 205
  difference, 22, 146, 165, 188n2, 210–11, 240
  divine, 50–1, 53, 56, 177, 179, 182
  divisibility of, 98–102
  in elemental change, 96
  of embryo, 222–3
  female contribution as, 3–4, 30, 34, 37, 41, 43–4, 72, 196, 207, 210, 221–2, 229, 236, 238
  form's work through, 12–13, 24, 49, 58, 147, 181, 187
  functional, 72, 88, 112, 239
  generative, 80–1
  generic, 88, 206
  independent character of, 23n60, 161, 204, 239
  indeterminate nature of, 23, 40, 236
  inert, 12, 34, 41–2, 47, 78, 81
  irrational nature of, 40, 42
  menses as, 4–5, 33, 59, 112, 144–6, 205, 235, 239
  moisture as, 154
  necessity, 63, 163–4
  opposed to form, 11, 42, 144
  passive nature of, 214
  as powerful, 48, 144
  as principle of individuation, 66, 68n58
  as privation, 205, 236
  proximate, 81–2, 112, 205
  restricted view of, 52–4, 62, 160, 170
  robust view of, 52, 55–8, 62
  in semen, 160, 176–9, 213–17
  as stuff, 2, 12, 42–43, 93, 108n88, 163–4, 236, 239
  as substratum, 86–92, 151
  suppression of, 10
Mayhew, Robert, 8, 11, 21n57, 33n29, 51n1, 188n1, 189n6
medical art, medicine, 12, 19, 40, 120–1, 138n, 214, 226, 234, 170
Melian nymphs, 124
menses, menstrual blood, καταμενία
  compared with semen, 22, 44, 49, 110, 114, 116, 141, 144–5, 159, 188, 190
  in craft view of generation, 3, 43, 77, 219
  Hippocratic view of, 138
  interaction with semen, 13, 19, 33, 63, 77–8, 80, 117–18, 160, 164, 168, 173, 179–81, 184–6, 191, 230–2, 238
  as matter, 5, 33, 59, 112, 144–6, 205, 235, 239
  in mythology, 126
  process of formation of, 8–9, 32, 58, 78, 119, 167, 183, 223, 238–9
  in resemblance, 195, 200, 202–3, 208
  as seed, 21, 31, 51n1, 86–7, 144
  in sexual differentiation, 210
menstrual blood *see* menses
Metis, 121, 129–30
Michael of Ephesus, 44
mind, 135
misogyny, 27–30, 33, 36, 42, 45
Möbius strip, 8, 13, 25, 77, 80, 118, 204, 210, 236, 240
moisture, moist
  in concoction, 146–60, 184
  elemental force of, 101, 104–9, 144, 220, 232, 239
  feminine association with, 9, 131, 137–40, 146
  in Greek myth and medicine, 12, 131
  Hippocratic view of, 132–4, 136–41
  Pre-Socratic views of, 132–7

moisture, moist (*cont.*)
   in semen, 55–6, 143, 180
   in sexual differentiation, 195
moles, 18
Monists, 98
monster, monstrosity, 33n28, 41, 143, 187–8, 188n3
mother, maternal
   generation occurs in, 216–17, 227
   Hippocratic view of, 138–9
   as material, 6–7, 10, 14, 24, 112, 162, 197, 219–20
   in mythology, 121–3, 125–31
   nutrients from, 166–8, 221
   positive contribution of, 46, 60, 170, 190, 198, 223
   resemblance to, 44–6, 197–9, 201–3
   separation from, 181
   in sexual differentiation, 191, 196
motion, movement
   in actualising a potential, 117
   artificial, 4, 48, 110, 172, 214–16
   elemental, 95–7
   of form, 62, 185, 213
   heat as cause of, 134, 154, 159, 173, 180, 183–4
   male as source of, 4–5, 32–3, 35–6, 52, 61, 72, 166, 177, 201, 213–22, 226
   natural, 47, 74, 94, 110, 146, 160, 172, 217, 238
   principle of, 4, 35, 109, 143, 170–1, 175, 194, 218–19, 220, 222, 232
   in resemblance, 5, 44–5, 197, 199, 201–3
   of semen, 32, 35, 55, 62, 77–8, 166–7, 170–2, 180–1, 194–5, 200–2, 208, 210, 216, 219, 224
   of soul, 4, 111, 166, 222–3
   vital heat as source of, 58–9, 183
moving cause, 4–5, 32–3, 61, 79, 114, 153, 162, 164, 170, 173, 214
mutilation, mutilated, 187–90, 193, 209

natural heat *see* vital heat
nature, φύσις
   craft model of 2n3, 2–5, 12, 15, 39–40, 47–9, 59n31, 79–80, 110, 118, 212, 215–26, 228, 241
   internal source of movement definition of, 1–3, 146, 161, 182, 220
   normative sense of, 37–8
   reliance on reason, 74
necessity, 41, 160–2, 236
   hypothetical, 163

Neoplatonism, 64–5, 70, 72, 75–6, 78, 81, 203, 218
non-homoeomerous, 43, 147, 159–62, 165, 168, 220
non-uniform *see* non-homoeomerous
Northwood, Heidi, 45, 198n29
Nussbaum, Martha Craven, 28, 177
nutrition, nourishment
   semen as actualising force of, 116, 193–4, 206–7
   semen and menses formed out of, 143–4, 164, 223
   as source of growth, 157, 167, 184, 192, 221, 234
   in stomach, 183
   vital heat as cause of, 57–8, 63, 139, 153, 166, 175, 182
nutritive soul, 4, 39, 55, 110, 162, 218, 221–3

Oedipus, 128
Okin, Susan Moller, 37
one-sex model, 16–22, 204, 238–41
   Aristotle as adherent of, 17, 20–2, 46–7
Orestes, 125–7, 129–30
organ, 17, 32, 58, 89, 112–13, 125, 147, 162, 167–8, 171, 174–5, 180, 192–3, 208, 226, 233
Ouranos, 122–123, 143
Owen, G. E. L., 73–4, 76

P-distinction, 73–4, 76
Pandora, 122–5, 131, 137–8
pangenesis, 34, 36, 169
pansomatism, 51, 53n8, 168–9, 225
Park, Katherine and Robert A. Nye, 20n52
Parmenides, 100–1, 135
patient, passivity, 5, 10–11, 45, 98–101, 104–5, 110, 114, 117, 213–14
Peck, 41, 198n29
Persephone, 130, 135
Phaethousa, 19
place, 95–6, 105, 130, 153, 160, 161n50
plants, 2, 4, 42, 84, 89, 124, 144, 150, 158, 167, 175, 183–4, 191, 221–2, 224–5, 228
Plato, Platonic, 6, 11, 23, 65, 68–76, 81–2, 91, 93, 94, 113, 140, 203
Platt, 41, 223
Plutarch, 126
pneumatisation, 32, 57
politics, political, 1, 5, 7, 13, 14, 17, 27–9, 31, 34, 37–8, 41–2, 44, 234–5
Posner, Richard, 20n52

potential, potentiality, 124–5, 149, 155, 172, 202, 205, 216, 223, 228
  active and passive, 45, 52, 116–18, 150, 116, 117n120, 118, 149, 171, 237
  artificial, 82
  capacity and exercise, 116, 119
  current and possible, 45, 119, 203
  female as, 15, 34, 43–4, 58, 58n24, 110, 188, 198, 221–2
  kinds of, 114–19
  knowledge, 66, 72, 76, 78, 115–17
  male, 44n73, 116
  matter as, 6, 10, 34, 36–7, 43, 48, 72, 77, 88–91, 94–7, 99, 101, 144, 160, 194–5, 225, 239
  movements, 45, 198n29, 201–3
  natural, 47, 90, 118, 220
  process and end, 116
  proper, 39
  pure, 38, 79–81, 89, 109
  in resemblance, 33, 173, 200–3
  semen as of soul, 169–70, 172, 183, 188, 198n29, 226
  in sex differentiation, 32, 58n24, 194
  soul, 111–14, 176, 226
potter, 216–17
Pre-Socratics, 48, 54, 57, 61, 104, 131–7, 141–2
preformationism, 28, 32, 36, 58, 169, 174
Preus, Anthony, 198n32
prime matter, 36, 41, 78, 79–83, 85–93, 97, 102–3, 107–9, 134, 212
principle, ἀρχή
  of elements, 148
  of form, 145, 172, 191
  heating, 137, 185
  of life, 121, 179, 191, 224, 228
  in the male, 193
  of movement, 1–5, 40, 53, 61, 192–4, 220, 232
  organising, 218, 220, 225
  privation, 46, 48, 73, 87–9, 97, 100–1, 115, 149, 153
Proclus, 74–6
Prometheus, 124
proper heat *see* vital heat
Ptolemy, 76
puppetry, puppets, 48, 58–9, 169, 219
Pythagoras, Pythagoreans, 42, 133
Pythia, 130–1

Rashed, Marwan, 178
rational soul, 6, 55–6, 72, 113–14
Reeve, C. D. C., 128, 198n29

relapse, λύεσθαι, 5, 44–5, 65, 187, 201–2
rennet, 184–5, 221, 227–32
resemblance, 5, 44, 65, 169
residue, 31, 61–2, 145–6, 152, 155, 160, 164, 174, 189, 192, 220–1
Rhea, 123
Robinson, H. M., 88n29, 90n33, 95n54
Ross, G. R. T., 182n114

sailor analogy, 111
Salamon, Gayle, 22
Salmieri, Gregory, 62n43
Scaltsas, Theodore, 68, 69n62, 83n9
Schleiner, Winfried, 18n45
Schofield, Malcolm, 9
seals, 188
Second Man Argument, 68–71
seed, σπέρμα
  male, 22, 29, 29n9, 32, 44, 56, 62, 124, 129, 138, 143, 205, 228
  menses as, 21n57, 30–2
  of plants, 89, 133, 221
  in sex determination, 44, 191
  source of embryo, 19, 36, 84–7, 89, 102, 108, 112, 210, 221
  sperm versus semen as, 51n1, 61, 86
  theories of, 21n57, 29–32, 139, 143, 194, 197
Semele, 122
semen, γόνη, 32, 51n1, 58, 143–4, 159–60, 164, 215
  cause of, 169–70
  definition of, 133, 145–6, 152, 155–9
  formation in father of, 165, 179
  heat in, 172, 183, 218, 220, 238–9
  menses' relation to, 112, 116–17, 206
  motion of, 55, 58, 62, 166, 171, 173, 218–19
  as moving cause, 165, 168, 170
  role of material in, 41, 51–6, 64, 110, 136, 141, 167, 170–1, 174, 176–9, 190–1, 200, 202, 217–20, 223
  in sexual differentiation, 193–7
  soul-causing role of, 36, 53–56, 59, 65, 72, 114, 166, 171, 174–6, 188, 200, 221–3, 239
  as tool, 199–200
sex differentiation, 5, 32, 45, 71, 176n89, 187, 191–7, 203
sexism *see* misogyny
sexual asymmetry, 14–15, 18, 20, 33
sexual difference, 5–6, 13, 16, 22, 31–2

shape, 163
Sharkey, Sarah Borden, 199n33
Shiffman, Mark, 185n134
Solmsen, Friedrich, 52–6, 77, 93, 181n112
Sophocles, 128, 136
soul
   as actuality, 72, 110–20, 176, 223, 232, 237
   form as, 6, 37, 40, 71–2, 74–5, 200
   heat's role in causing, 63, 133, 150, 165–6, 175–7, 179–80, 181n112, 182–3, 185, 222, 225, 237
   Hippocratic view of, 133–4, 180
   male as source of, 3–4, 17, 34–6, 47, 55
   material basis of, 53–61, 77, 110, 161–2, 181, 218, 236
   Pre-Socratic views of, 134, 136
   relation to body of, 3–4, 25, 36, 72, 110–14, 116, 118, 167–8, 176, 200, 223, 226–7, 237
   semen as, 41, 50, 72, 144, 155, 160, 165–8, 170–1, 173–4, 176, 188, 200, 220, 226, 232, 236, 239
soul heat *see* vital heat
sperm *see* semen
Stocks, J. L., 92
Stoics, 61
stomach, 57, 124–5, 133, 141, 183–6, 235
Studtmann, Paul, 57, 185
substance, 4
   natural, 70n65, 79, 81–92, 102, 104, 109, 152, 173, 183, 197, 238
   relation of form to, 64–73, 75–8
   as substratum, 90–2
   unity of, 24
substratum, ὑποκειμένων, 69–70, 73, 80, 82–92, 101, 106–9, 135, 150–5
sun, 50–1, 56–7, 134, 137, 156, 179, 181–3, 227–8
sword, 172, 220
Sybil, 131

teacher, 216
teleology, 37–40, 43, 46, 48, 62–4, 161–2, 179, 191
   primary, 164–5
   secondary, 164
testacea, 224–5

testes, 192
Thales, 133–4, 177, 181
theological vitalism, 54, 60
tool, tools
   body as, 112–14, 168
   form as, of efficient cause, 153
   heat as, 76, 161, 173
   male as, 35
   material as, 77, 163, 225
   semen as, 61, 110, 114, 144, 171–2, 199–200, 215–20, 222, 236
   soul working through, 3, 4, 113–14, 118, 181n112, 223, 225–6, 229, 237
Tress, Daryl McGowan, 8–9, 30–1, 34–7, 49, 53, 53n8, 61, 77, 181n111, 212, 212n2
Tuana, Nancy, 127–8
two-sex model, 17–18, 77, 204, 238–41
   Aristotle as adherent of, 20–2, 47
Typhon, 123

uniform *see* homoeomerous
uterus, 191–2, 221

vital heat
   animating capacity of, 134, 177, 181n112, 224–5
   contrasted to fire and heat, 53, 56–7, 63, 175, 179–80, 220, 228, 230
   as difference between male and female, 29n8, 32, 41, 144, 146, 148, 155–7, 207
   distinguishing between kinds of soul, 57, 185
   role in semen, 50–6, 110, 144, 165

water, 41, 89, 94, 96, 101, 103, 106–9, 132–3, 135, 138, 156, 178, 224–5
wax, 213–14
Webster, E. W., 150n19, 153n31
wet, 89, 101, 132, 136, 140–1
whales, 189
Whiting, Jennifer E., 67–8
Williams, C. J. F., 97n59
Witt, Charlotte, 11, 37–8, 45, 66, 71, 116, 119, 198n32, 199n33
Woods, Michael J., 64n48, 69n63

Zeller, 88n29
Zeus, 121–4, 128–31, 135, 137

EU representative:
Easy Access System Europe
Mustamäe tee 50, 10621 Tallinn, Estonia
Gpsr.requests@easproject.com

www.ingramcontent.com/pod-product-compliance
Lightning Source LLC
Chambersburg PA
CBHW082141230426
43672CB00016B/2930